AN INTRODUCTION TO NETWORK MODELING AND SIMULATION FOR THE PRACTICING ENGINEER

A volume in the IEEE Communications Society series:
The ComSoc Guides to Communications Technologies

Nim K. Cheung, *Series Editor*
Thomas Banwell, *Associate Editor*
Richard Lau, *Associate Editor*

Next Generation Optical Transport: SDH/SONET/OTN
Huub van Helvoort

Managing Telecommunications Projects
Celia Desmond

WiMAX Technology and Network Evolution
Edited by Kamran Etemad, Ming-Yee Lai

An Introduction to Network Modeling and Simulation for the Practicing Engineer
Jack Burbank, William Kasch, Jon Ward

AN INTRODUCTION TO NETWORK MODELING AND SIMULATION FOR THE PRACTICING ENGINEER

Jack Burbank
William Kasch
Jon Ward

The ComSoc Guides to Communications Technologies
Nim K. Cheung, *Series Editor*
Thomas Banwell, *Associate Series Editor*
Richard Lau, *Associate Series Editor*

IEEE PRESS

A JOHN WILEY & SONS, INC., PUBLICATION

Published by John Wiley & Sons, Inc., Hoboken, New Jersey.
Published simultaneously in Canada.

For general information on our other products and services or for technical support, please contact our Customer Care Department within the United States at (800) 762-2974, outside the United States at (317) 572-3993 or fax (317) 572-4002.

Wiley also publishes its books in a variety of electronic formats. Some content that appears in print may not be available in electronic formats. For more information about Wiley products, visit our web site at www.wiley.com.

Library of Congress Cataloging-in-Publication Data is available.
ISBN: 978-0-470-46726-8

oBook ISBN: 978-1-118-06365-1
ePDF ISBN: 978-1-118-06363-7
ePub ISBN: 978-1-118-06364-4

Printed in Singapore.

10 9 8 7 6 5 4 3 2 1

■■■■ CONTENTS

This book provides an overview of the current state-of-the-art in modeling and simulation (M&S) tools and discusses many of the pitfalls most commonly encountered by network engineers. A bottom-up approach is taken in describing network M&S, following the Transport Control Protocol / Internet Protocol (TCP/IP) modified Open System Interconnect (OSI) stack model. While applicable to network M&S in general, there is particular emphasis placed on wireless network M&S. This book first decomposes the wireless network M&S problem into a set of smaller scopes: 1) radio frequency (RF) propagation M&S (Chapter 2), 2) physical layer (PHY) M&S (Chapter 3), 3) Medium Access Control (MAC) layer (Chapter 4), and 4) higher layer M&S (Chapter 5). After considering each of these smaller scopes somewhat independently, the book then revisits the overall problem of how to conduct M&S of a wireless networking system in its entirety.

No specific assumptions are made on the type of network being modeled in any particular layer of the protocol stack. Instead, the building blocks are presented to address the common challenges of modeling any wireless network. The reader is also directed to resources that provide more detail on specific topics. Resources are chosen from generic studies of wireless networks and from the Mobile Ad Hoc Network (MANET) and ad hoc sensor network communities. This book is written with particular emphasis placed on specific topics at the different layers of the protocol stack, with the intention of bridging gaps between the computer science and electrical engineering communities. Historically, the higher layers of the protocol stack are often considered research subjects for computer scientists and the lower layers for electrical engineers. In fact, accurate simulations must capture the cross-layer interactions and higher layer simulations must consider the impacts of the lower layer conditions on results. The authors hope that this book will educate the reader in simulation topics that may have not otherwise been considered and will ultimately lead to improved simulation results in the wireless networking research community.

This book can improve the reader's background knowledge on the key components of successful wireless network simulations. But, ultimately, the reader must learn to validate his or her own simulation since they alone will know all specific details and assumptions that lead to a specific result. In general, the output of a simulation should not be a surprise to the designer, and, it if is, sufficient research into the underlying protocol must be conducted to explain any unanticipated results. Because there are so many variables present in a model and therefore so many potential locations where errors are introduced, a model output should not be taken as ground truth without other methods of verification. Results may be compared with results from other researchers, but as some papers [1–4] note, results between two equivalent scenarios simulated on two different simulators may not match. In this case, the designer must not only validate whether or not his or her simulation is correct, but also what led to results not matching the other simulation. Results should not be published until the simulation designer has confidence in the model, the results have been validated to the best of the designer's ability, and, once published, should contain all model parameters, assumptions, and simulation source code.

In this book only a select set of simulators have been considered as the most popular commonly used by academic and industrial researchers. These include OPNET, NS-2, GloMoSim, and QualNET. There is no single, all-purpose simulator that is best for all scenarios. Additionally, budget constraints often force researchers to choose open-source simulators over commercial solutions. Custom simulation solutions (i.e., homebrew simulations) are certainly too numerous to be considered. Note that the risk of citing specific simulators is that these tools are continually evolving. This means that statements about a given product's current capabilities may no longer be valid, as subsequent releases enhance a tool's capabilities. Care has been taken by the authors to focus on principles and practices that assist the simulation designer in improving wireless network simulations while remaining independent of a particular simulator, and hence topics and results are not as limited to an expiration date.

JACK BURBANK
WILLIAM KASCH
JON WARD

To view color versions of the figures in this book, please visit http://booksupport.wiley.com.

ACKNOWLEDGMENTS

We would like to acknowledge the numerous individuals who have helped make this book a reality. First and foremost, we would like to acknowledge Brian Haberman and Julia Andrusenko for their assistance in writing this book, contributing their expertise and understanding of network simulation tools and radio frequency propagation tools, respectively.

We would like to thank Robert Nichols for his long-time support of our activities in this field.

We would like thank all of our friends and family for their patience and support during the writing of this book.

ABOUT THE AUTHORS

Jack L. Burbank (jack.burbank@jhuapl.edu) received his B.S. and M.S. degrees in electrical engineering from North Carolina State University (NCSU) in 1994 and 1998, respectively. As part of the Communications and Network Technologies Group of The Johns Hopkins University Applied Physics Laboratory (JHU/APL), he works with a team of engineers focused on assessing and improving the performance of wireless networking technologies through test, evaluation, and technical innovation. His primary expertise is in the areas of wireless networking and modeling and simulation, focusing on the application and evaluation of wireless networking technologies in the military context. He has published numerous technical papers and book chapters on topics of wireless networking, and regularly acts as a technical reviewer for journals and magazines. He teaches courses on the topics of networking and wireless networking in the Johns Hopkins University Part Time Engineering Program, and is a member of the IEEE and the ASEE.

William T.M. Kasch (William.kasch@jhuapl.edu) received a B.S. in electrical engineering from the Florida Institute of Technology in 2000 and an M.S. in electrical and computer engineering from Johns Hopkins University in 2003. His interests include various aspects of wireless networking technology, including MANETs, IEEE 802 standards, and cellular. He participates actively in both the IEEE 802 standards organization and the Internet Engineering Task Force (IETF).

Jon R. Ward, PE (jon.ward@jhuapl.edu) graduated from NCSU in 2005 with an M.S. degree in electrical engineering. He works at JHU/APL on projects focusing on wireless network design and interference testing of standards-based wireless technologies such as IEEE 802.11, IEEE 802.15.4, and IEEE 802.16. He has experience in wireless network modeling and simulation (M&S) and test and evaluation (T&E) of commercial wireless equipment. He is currently a student at the University of Maryland, Baltimore County (UMBC), pursuing a Ph.D. degree in electrical engineering.

CHAPTER 1

Introduction

Communications systems continue to evolve rapidly. Users continue to demand more high-performance networking capabilities. Service providers respond to this demand by rapid expansion of their network infrastructure. Network researchers continue to develop revolutionary new communications techniques and architectures to provide new capabilities commensurate with evolving demands. Equipment vendors continue to release new devices with ever-increasing capability and complexity. Technology developers rapidly develop next-generation replacements to existing capabilities to keep up with demand. These rapid developments in the network industry lead to a large, complex landscape.

The network designer and developer wants (and needs) to satisfy the demands of the users. This is difficult, as it is often complicated for the typical network engineer to fully understand this rapidly evolving communications landscape. This challenge is exacerbated by the nature of emerging technologies and techniques that are often extremely complex compared with their legacy counterparts. This leaves the typical network engineer with more questions than answers. The network engineer tasked with maintaining an operational network might ask the following: What is the right approach to solving my problem? Do I buy the latest device from company X that claims to solve all my problems? Do I replace the underlying technology of my system with the latest generation? How do I know whether a technology is mature enough to survive the rigors of my application? How do I know how my already existing network system will respond if I add this device? The network engineer researching next-generation networking techniques might ask: How do I know how this new approach will interact with already-existing protocols? or How do I build confidence in the utility of this approach without producing and deploying the technology? The network engineer developing a particular product might ask: How do I ensure that this design will satisfy requirements

An Introduction to Network Modeling and Simulation for the Practicing Engineer, First Edition.
Jack Burbank, William Kasch, Jon Ward.
© 2011 Institute of Electrical and Electronics Engineers. Published 2011 by John Wiley & Sons, Inc.

before I go to production? or How can I assess the utility of a design choice compared to its envisioned cost? This book aims to help answer these questions.

There are many tools available to the network engineer that can assist in answering these questions, including analysis, prototype implementation and empirical testing, trial field deployments, and modeling and simulation (M&S). It should be stated now that no one tool is typically sufficient in understanding the performance of a network; unfortunately, there is no "silver bullet" answer to all our questions. The complex nature of emerging systems also introduces significant complexity into the effective evaluation of these systems and how these various tools can be employed. Evaluation is often conducted through the coordinated usage of analysis, M&S, and trial deployments in closely monitored environments. Due to the costs and complexities of deployments, analysis and M&S are often used to determine the most sensitive performance areas that are then the focus of trial deployments. This limits the scope of the trial deployment to a realistic level while focusing on the important cases to consider.

Because of the increasingly interconnected nature of communications systems, and the resulting interdependencies of individual subsystems to operate as a whole, it will often be the case that individual subsystems cannot be tested in isolation. Rather, multiple systems must be evaluated in concert to verify system-level performance requirements. This increases the required scale of trial deployments and adds significant complexity as now several different types of measurements will often be required in several different locations simultaneously. This increases the required support for a deployment in terms of required resources, including personnel and measurement equipment, further limiting the realistic amount of trial deployments. Thus, this will place a premium on analysis and M&S to perform requirements verification and to form the basis of any performance evaluation. In many cases, M&S may provide the only viable method for providing insight into the behavior of the eventual system prior to full-scale deployment.

Once the importance of M&S is established, many additional questions still arise: How does the network engineer properly employ M&S? What are the most appropriate M&S tools to employ? While networking technologies continue to evolve rapidly, so too do M&S tools intended to evaluate their performance. The M&S landscape is indeed a complicated space with a multitude of tools with a variety of capabilities and pitfalls. Furthermore, there is often a poor understanding of the proper role and application of M&S and how it should fit within the overall evaluation strategy. There is even confusion surrounding the term M&S itself. Before we continue, let us provide some basic definitions that will be used throughout the book.

Modeling and simulation (M&S) are often combined as a single term. However, a model is quite different than a simulation. This book defines these two entities as:

Model: A logical representation of a complex entity, system, phenomena, or process. Within the context of communications and networking, a model is often an analytical representation of some phenomena (e.g., a mathematical representation for the output of a system component) or a state machine representation. This analytical representation can either be in a closed form or an approximation obtained through assumptions.

Simulation: An imitation of a complex entity, system, phenomena, or process meant to reproduce a behavior. Within the context of a communications network, a simulation is most often computer software that to some degree of accuracy functionally reproduces the behavior of the real entity or process, often through the employment of one or more models over time.

Emulation: An imitation of a real-world, complex entity or process meant to perfectly reproduce a behavior or process. Emulation can be thought of as perfect simulation of something such that it is equivalent to the original entity.

To illustrate the difference between a model and a simulation, consider a simple signal detection circuit. A simulation of this device would imperfectly mimic the various actions of the detection circuit to determine a likely outcome for a given input. A model of this same device would generally take the form of a mathematical algorithm that would produce (either perfectly or imperfectly) an output for a given input.

Unfortunately, the terms *model* and *simulation* are often incorrectly used interchangeably. Generally speaking, the term simulation has wider scope than the term model, where a simulation is typically a compilation of models and algorithms of smaller components of the larger overall entity or process. This book generally uses the combined term *M&S* to generically refer to the employment of models, simulations, and emulators to approximate the behavior of an entity or process.

There are numerous types of computer models and simulations. A computer model or simulation can generally be classified according to several key characteristics:

- Stochastic vs. Deterministic: Deterministic models are those that have no randomness. A given input will always produce the same output given the same internal state. Deterministic models can be defined as a state machine. Deterministic models are the most common type of computer model. A stochastic model does not have a unique input-to-output mapping and is generally not widely employed, as it leads to unpredictability in execution. A simulation can be made to act in a pseudo-random manner through the employment of random number generators to represent random events. However, the particular models governing the behavior of each component within the simulation are generally deterministic.

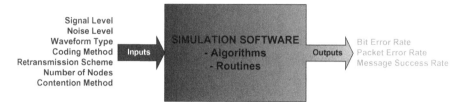

FIGURE 1-1. A block diagram of a wireless communications system simulation.

- Steady-state vs. Dynamic: Steady-state models attempt to find the input-to-output relationship of a system or entity once that system is in steady-state equilibrium. A dynamic simulation represents changes to the system in response to changing inputs. Steady-state approaches are often used to provide a simplified model prior to dynamic simulation development.
- Continuous vs. Discrete: A discrete model considers only discrete moments in time that correspond to significant events that impact the output or internal state of the system. This is also referred to as a discrete-event (DE) model or DE simulation. This requires the simulation to maintain a clock so that the current simulation time can be monitored. Jumps between discrete points in time are instantaneous; nothing happens between discrete points in time corresponding to interesting events. Continuous simulations consider all points in time to the resolution of the host's hardware limitations (all computer simulations are discrete to some extent because of the fact that it is running on a digital platform with a finite speed clock). DE methods are the most commonly used for network M&S.
- Local or Distributed: A distributed simulation is such that multiple computer platforms that are interconnected through a computer network work together, interacting with one another, to conduct the simulation. A local simulation resides on a single host platform. Historically, local simulations have been the most common. But the increasing complexity of simulations have increased the importance of distributed simulation approaches.

In general, a simulation can be thought of as a piece of software residing on a computer platform that implements a set of algorithms and routines and takes a set of inputs to produce a set of outputs that represent the behavior of the system of interest. This is depicted in Figure 1-1.

The typical inputs that are important to consider when simulating a wireless network are summarized in Table 1-1. The typical outputs that are often of interest are summarized in Table 1-2.

TABLE 1-1. Typical Inputs to a Wireless Network Simulation

Parameter	Explanation
Signal power	This will influence the received power level and consequently the Bit Error Rate (BER) and Packet Error Rate (PER) performance of the wireless link.
Waveform type	This will influence the BER and PER performance of the wireless link in a given channel.
Forward error control coding (FEC) method	This will influence the BER and PER performance of the wireless link in a given channel.
Retransmission protocol	This will affect the throughput and delay performance of the wireless link.
Contention method	This will influence BER, PER, throughput, and delay performance of the wireless link in a given channel.
Channel model	This will determine the performance of a given wireless link in terms of received power level, BER, and PER.
Mobility model	This will impact the performance of the MAC layer protocol and of the higher layers (e.g., IP routing).
Traffic model	This will impact the performance of the MAC layer protocol and of the higher layers (e.g., IP routing).
Network topology	This will impact the performance of the MAC layer protocol and of the higher layers (e.g., IP routing).

TABLE 1-2. Typical Outputs from a Wireless Network Simulation

Parameter	Explanation
BER	The fundamental performance metric of a digital communications link.
PER	Often considered the most important performance metric in a packet-switched network.
Throughput	The data rate supportable by the wireless network.
Goodput	The useful data rate supported by the wireless network (i.e., data rate as available by the application).
Latency	The end-to-end delay that an application or user will experience across the wireless network.

1.1 ADVANTAGES AND DISADVANTAGES OF MODELING AND SIMULATION

As is the case with any tool, M&S has both advantages and disadvantages. This section provides a tradeoff framework for the designer or developer to consider when choosing to employ M&S. In the following section, M&S is often compared with empirical testing. For the purposes of this book, empirical testing refers to real-world testing of equipment (e.g., physical hardware devices) deployed in a physical environment.

1.1.1 Breadth of Operational Scenario

First and foremost, M&S provides the ability to exercise a wide range of operational scenarios. Empirical testing will exercise a much smaller portion of the possible scenario space than will M&S. This includes the ability to evaluate greatly increased network scale (e.g., number of network nodes), not easily achieved in empirical activities, and more dynamic choice of environmental conditions (e.g., wireless environment). Because of the ability to exercise a wide variety of scenarios, M&S has a clear advantage in this aspect.

1.1.2 Cost

Generally, another advantage of M&S is reduced cost compared with empirical testing and trial deployments. Extensive empirical testing carries a high cost, to the point where extensive empirical-only approaches are largely impossible in the modern wireless networking landscape; however, this advantage is dependent on the scope placed on the M&S development effort.

1.1.3 Confidence in Result

A less obvious advantage of M&S is the amount of precision and control that can be exerted over the scenario in question. In the empirical scenario, measurements are taken and then those measurements are analyzed and understood for their ramifications. However, due to the uncontrolled nature of empirical testing, there are often many variables that affect the measurement. And often the number of uncertain variables is so great that it is impossible to isolate the source of any behavior or to correlate a measurement to its source (i.e., map the effect to the cause). This limits the scientific utility of such measurements, and makes it difficult to associate a high degree of confidence to the measurement. The "the data is what it is" philosophy is rarely justified if the phenomena under observation are not understood. Note, this is much more the case for over-the-air (OTA) empirical activities. Other empirical activities are much more highly controllable (e.g., direct radiofrequency (RF) chain testing).

The primary, and most obvious, disadvantage of M&S is that it is not real. It is a representation of the system, rather than the system itself. There are several assumptions that will be built into any M&S tool. Some of these assumptions will be necessitated by real-world complexities that are not easily represented. Others are necessitated by a lack of information available about the system in question. This will naturally lead to inaccuracies. Consequently, this leads to a decreased confidence in results. This confidence decrease is manageable, however, through verification and validation activities, often in conjunction with empirical activities to improve confidence in such models.

A higher degree of confidence is almost always associated with empirical methods, regardless of the methodology or practices employed during those empirical activities. Unfortunately, this confidence can be ill placed. The common belief is that M&S-based methods are more subject to error because software-based "bugs" could introduce unforeseen inaccuracies. And while that is definitely true, the same applies to the empirical-based approach. Any empirical measurement will have error associated with it (e.g., imperfections in hardware employed to make a measurement, misconfiguration of test equipment). Also, human interpretation must at some point be applied to understand an empirical measurement. This human interpretation can be influenced by assumptions, biases, and preconceived opinions.

Another issue is that of statistical significance. Even if measurement error has been minimized, there are several factors that can influence the significance of that measurement. Take, for example, the measurement of an antenna pattern, which is a key characteristic that will impact wireless network performance. This antenna pattern will vary across antenna population due to manufacturing variation, differences in platform, and differences in age and condition. Furthermore, the RF propagation environment characteristics will be temporal in nature. Thus, a particular measurement is somewhat insignificant in the overall sense. In fact, to make empirical activities truly significant from a statistical standpoint is often cost prohibitive.

With all these factors considered, an empirical approach is still considered to have an advantage, especially if issues such as measurement error and uncertainty are built into empirical activities. However, the proper application of verification and validation practices can help minimize this difference.

1.1.4 Perception

Even if a model is highly accurate, and from a scientific perspective is highly regarded, there is the issue of perception. Many individuals will still remain skeptical of the results from a computer model. This is due to sociological and psychological phenomena that are well beyond the scope or timeframe of any particular M&S activity. Rather, this reality must be accepted and factored into the overall evaluation approach. An empirical-based evaluation method has the overwhelming advantage in this area. In fact, this advantage is so strong that some degree of empirical testing is likely required to give credibility to the findings of the overall M&S activity.

1.1.5   The Need for Verification and Validation

While not considered a disadvantage, certainly a burden associated with M&S is the need to conduct verification and validation (V&V) activities. Such activities are generally required to both verify the accuracy and consistency of model output and validate output relative to other models, empirical tests, and theory. While V&V activities are mandated by good software engineering principles and must be adhered to, the formality of a V&V process can levy significant resource requirements on a project. This partially negates the cost advantage of M&S over empirical testing.

In some sense, M&S is disadvantaged in this regard compared with other tools available to the network engineer. As mentioned previously, there is typically less scrutiny placed on empirical measurements and, consequently, there is typically a greater "burden of proof" placed on an M&S developer as compared with the empirical tester.

1.2   COMPARISON OF "HOMEBREW" MODELS AND SIMULATION TOOLS

Custom simulations, or "homebrew" solutions, are those in which the implementer does not rely on any existing tools but rather develops the simulation in its entirety. The advantages of homebrew simulations include:

- The implementer knows exactly what has been implemented.
- Homebrew solutions can have significant performance benefits.

The disadvantages of homebrew simulations include:

- They can be costly to develop.
- They can be difficult to upgrade.
- There is a real risk of these custom simulations not being widely adopted, even within your organization (resulting in perpetual "homebrew" solutions).

Other than small-scale efforts that are supporting analysis, homebrew approaches are generally discouraged. With the ever-increasing complexity of wireless networking systems, the feasibility of a meaningful homebrew solution is dwindling. Even for cases where there are no existing implementations of a particular networking technology and code development is inevitable, it is recommended that this new custom simulation be developed within existing tools/environments so that it can be integrated with and leverage existing simulation libraries.

1.3 COMMON PITFALLS OF MODELING AND SIMULATION AND RULES OF THUMB

There are many potential pitfalls that face those who embark on a network simulation development effort. This section discusses some of those most commonly seen.

1.3.1 Model Only What You Understand

It can be said that the utility of a given model is only as good as the degree to which it represents the actual system being modeled. Indeed, a system—whether a wireless network or otherwise—can only be modeled once it is sufficiently understood. While this is a simple tenant, it is one that is certainly not adhered to universally by M&S designers. One may ask why M&S designers develop invalid models. There are many reasons, the first of which is that high-fidelity model development requires a significant investment of time and effort. This statement is not meant to offend developers or to imply carelessness on their part. The fact is that many designers are under time constraints to deliver results. Consequently, a careful understanding of the underlying system being modeled and rigorous validation of the model is not always an option.

While understandable, this is at the same time unacceptable. It is highly unlikely that a simulation developer can provide a meaningful result when they did not understand the system they were intending to model. While the timeline might have been met, the result was likely meaningless. Worse yet, the result was likely wrong and might have adversely affected larger design or business decisions. *Model only what you understand!* If you don't have a fundamental understanding of a technology, there is no way you can effectively model or simulate that technology. This step cannot be skipped in a successful M&S effort. If this step cannot be completed, it is better to not proceed down the path of M&S development.

1.3.2 Understand Your Model

It is quite common for the network engineer to utilize off-the-shelf tools, either commercial or open source. This approach typically lends itself to a faster M&S development cycle; however, it is imperative that the network engineer has a full understanding of the tools being used. Most simulations are likely to have errors—even commercial tools. New simulation implementations almost always contain errors. Simulation implementations can make assumptions that may not accurately reflect the exact performance metric of interest.

If the simulation developer utilizes existing simulation implementations, it is imperative to allocate the proper amount of time to closely examine that implementation to fully understand what that code is doing and what it is not doing. There is no better way to lose credibility than to not be able to answer questions about one's own results. *Understand what you have modeled!* There are resources available to help with this, including technical support for commercial tools, online newgroups and user forums for open source tools, and in some cases the simulation designer can contact the author directly (e.g., a contributed simulation to an open source project).

1.3.3   Make Your Results Independently Repeatable

Many academic papers such as [1–3] have discussed the lack of independent repeatability in wireless network simulation results due to improper documentation of the simulator being utilized, model assumptions, and inputs and outputs. There are subtle parameters and assumptions embedded in simulators such as NS-2 and GloMoSim that certainly can impact all results. Often default simulator parameters are chosen that may not capture the intended network conditions for a given scenario [2]. Perhaps the larger problem is that simulation results are often presented as ground truth and not as a relative ranking of a new idea compared to existing ideas. That is, the literature survey component must always be present in wireless network research and simulation results should be compared to existing results to demonstrate advantages and disadvantages of new ideas. Moreover, new simulation results must be compared with results in existing literature using the same simulator, underlying assumptions, and parameter conditions.

1.3.4   Carefully Define M&S Requirements

This is an activity that is too often ignored or given superficial treatment. The authors would argue that network engineers all too often rush into an M&S effort without a clear idea of what they are hoping to accomplish. This is a surefire recipe for failure.

The first step is to clearly understand the metrics of interest that would be generated by a simulation. Is overall network throughput the metric of interest? Is BER the metric of interest? End-to-end delay? Not all simulation tools necessarily lend themselves to the same types of output metrics, so it is important to define these metrics so that tool selection is an informed process.

The next step is to clearly define the required performance of the simulation to be developed. This book contends that there are four primary dimensions of performance:

- *Cost*: The overall investment in resources towards the development and maintenance of the M&S activity. This includes not only original platform costs, but also development time, upgrade and maintenance costs, and troubleshooting.

- *Scalability*: The total complexity of the system to be simulated. There are two factors that must be considered: network size in terms of number of nodes, and network traffic model in terms of number of messages per unit time. These two factors will drive the computational complexity of the simulation and will ultimately be the limiting factors in the size of the network that can be simulated. This is generally governed by software complexity and hardware capability.

- *Execution Speed*: For a given simulation scenario, how quickly can that simulation complete and provide the desired output metrics? This is generally governed by software complexity and hardware capability.

- *Fidelity*: For a given simulation scenario, how accurately does the simulation's output metrics reflect the performance of the real system.

Note that these dimensions of performance are contradictory; not all performance dimensions can be achieved simultaneously. If you desire a highly scalable simulation with fast execution speed, then the fidelity is likely going to be lower. Do you want high fidelity and scalability with reasonable execution speed? Then the cost will likely be very high. In general, you can pick any three of these metrics.

A common pitfall is to begin an M&S effort with unrealistic expectations. Is it really feasible to model the entire Internet down to every platform with bit-level fidelity? Probably not. Is it possible to model the entire Internet down to every platform with many simplifying assumptions? Probably, but it is unlikely to be useful.

When defining requirements and expectations for an M&S effort it is recommended to begin by choosing the required fidelity. How accurate of an output metric is required? A successful effort will always begin with this metric because, without a meaningful degree of fidelity, any M&S activity is meaningless, despite its scalability or execution speed. Once the required fidelity is established, one can then begin placing limitations on simulation capabilities accordingly. Cost is generally bound by an allocation of resources. So given a known cost constraint and a known fidelity requirement, we can then begin building a conceptual model for the simulation. The target fidelity will mandate the inclusion of particular system characteristics with great detail and inputs with particular degrees of accuracy, and also allow for relaxation on other system details and input accuracy. Note that this exercise requires a strong understanding of the system being modeled and on the underlying concepts of wireless networking. Remember, model only what you understand! Once a conceptual model is designed, the hardware platform can be chosen in accordance with cost constraints to maximize scalability and execution speed performance.

1.3.5 Model What You Need and No More

One of the first decisions that the simulation designer must face is to determine what he or she is attempting to demonstrate through simulation and what is the most simplistic model that captures all necessary components. The engineering tradeoff is that increased detail can provide higher-fidelity output from the model, but at the cost of complexity—potentially introducing error and certainly increasing debugging time and execution time. The designer must also realize that a model is always an abstraction from the real world. Wireless networking devices not only have variables within the standards to which their underlying protocols comply, but there is variability introduced into each manufacturer's products. At least a subset of the key variables should be included: transmission power, antenna type and gain, receiver sensitivity, and dynamic range should be considered in the model, but the extent of modeling detail required depends on the particular system and desired output for a given scenario. Regardless of the level of detail included, a simulation will always be an approximation of the real system; an arbitrarily high degree of fidelity is generally not possible. Also, the cost of increased fidelity at some point becomes greater than the marginal utility of the additional fidelity. This is illustrated in Figure 1-2. It is imperative to understand the limitations of M&S techniques and to understand the relationship between cost and fidelity so that an M&S effort does not become an over-engineered effort in futility.

How much detail is sufficient in a simulation to capture the essence of the real-world network being modeled? Unfortunately, the answer to this question is that it depends on the particular simulation scenario. The reader should first decide exactly what is the problem that he or she seeks to address through simulation. What are the inputs and the outputs of the model? Some outputs may be independent of specific details in the model, while others may be correlated and therefore seriously affected if those components are abstracted.

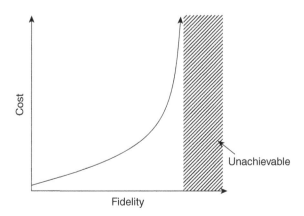

FIGURE 1-2. The cost of simulation fidelity.

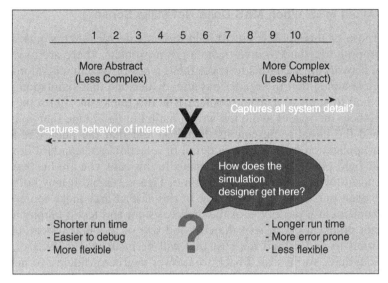

FIGURE 1-3. Illustration of the simulation trade space.

Simulation always takes the form of an abstraction of a system to allow the designer to gain some insight from investigating various operating scenarios of the system. In many cases the simulation allows the user access to knobs and switches that may not be available on the actual system. Consider reliability testing for a consumer networking product that must be tested under as many operating conditions as possible, where the prevention of erratic behavior in a consumer product translates to significant savings for a company. Yet in other cases the researcher desires to investigate a system's reaction to a single condition that may be unlikely to occur in real life. Perhaps testing the actual system under this condition could be harmful and simulation is the only way to examine the problem.

The next step is to decide how much of the system must be implemented for the simulation results to be valid. Ultimately, the reader is going to have to decide the level of detail required in his or her simulation, but this book is intended to guide the reader towards formulating a more educated decision. First, the reader must consider the engineering tradeoffs between adding more detail to a model and increased computational time, increased complexity, and increased debugging time. This simulation trade space is illustrated in Figure 1-3. A more complex simulation may attempt to capture an actual system's complete behavior, but at that point the simulation is generally inflexible to scenario modifications, more prone to errors, and more computationally intensive. A more abstract approach that focuses only on the basic behavior of a system is generally very flexible, easier to debug, and has a shorter execution time. But, it may not capture the behavior of interest.

1.3.6   Avoid M&S When M&S Does Not Make Sense

The purpose of this book is not to help the user decide when simulation is the appropriate method to investigate a given problem. There are too many possible networking scenarios to make these types of recommendations. It is therefore assumed that the reader has already decided that simulation is the best method to apply to a given problem. But, this book will offer some basic advice to help the reader avoid the wrong path. Let us assume that you have performed the initial requirements definition and that the fidelity required for your application includes every detail of a technology standard down to every bit, byte, protocol, and state machine. In this case, you are likely on the upper end of the cost vs. fidelity function of Figure 1-2 and it may not make sense to even pursue an M&S activity. In this case, it may make more sense to just implement a prototype of the device/system and test it empirically. If that is not practical because of the cost and size of the final system, then it is important to understand the cost that will be incurred by M&S, or such efforts may have to be scaled back to a lower degree of fidelity to manage cost. It is ultimately up to the reader to decide if M&S is right for their particular effort.

1.3.7   Channel Models

A quick search of open literature will uncover a plethora of highly complex models of wireless networks and proposed protocols/techniques that are evaluated only in Additive White Gaussian Noise (AWGN) environments (none are referenced here to protect the names of the innocent). This is perfectly fine for many cases. With that said, however, do not expect to model an omni-directional antenna wireless network in AWGN conditions and be able to make any statements regarding how that system will behave in complex urban environments. It is sometimes a daunting task to provide high-fidelity channel models in large simulations. This is well understood. But it is important to understand and clearly communicate the limitations of the model to constrain performance statements, particularly if those performance statements are going to form the basis for design or business choices. Common RF channel models are discussed in detail in Chapter 2.

1.3.8   Mobility Models

There are many papers in open literature that present the types of mobility models to use when simulating wireless networks (e.g., [5]). However, it is important to understand that, while mobility models will have a profound impact on the performance of the network, they are usually arbitrary and hardly ever reflect reality. It is indeed difficult to predict the true mobility patterns of network users, particularly future patterns. It is important for the simulation designer to do his or her homework and construct the best educated guess when formulating mobility models for use in simulations. It is also important to perform sensitivity analysis to understand how the metrics of interest change with different mobility models to understand the M&S limitations for a particular application. Simplistic assumptions combined with the lack of expectation management can (and usually will) haunt you!

1.3.9 Traffic Models

Like the case of mobility models, traffic models usually have a profound impact on the performance of a network. And, unfortunately, like the case of mobility models, traffic models are also usually arbitrary and hardly ever reflect reality. It is generally possible to construct realistic current traffic models based on traffic monitoring and analysis. But in the case of a new network deployment, it is difficult to ascertain the true pattern of usage. And it is also very difficult to predict future usage patterns since applications evolve rapidly. It is important for the simulation designer to do his or her homework and make the best educated guess possible. However, be cognizant that these are still guesses, best case. It is also imperative to perform sensitivity analysis to understand how the metrics of interest can change with changes in traffic patterns to understand the M&S limitations of a particular application. Again, simplistic assumptions combined with the lack of expectation management can (and usually will) haunt you!

1.3.10 Over-reliance on Link Budget Methods for Abstraction

Even in simulation environments, it is common to simplify complex aspects of the system and turn them into static "losses" in link budgets (e.g., signal quality adjustments at a receiver to represent some physical phenomena causing degradation). This is fine for a simple, steady-state analysis. But in the more general dynamic case, beware that losses are typically scenario dependent. In this case, it is important to understand the degradation source and its sensitivity to scenario-dependent variables. Once sensitive variable relationships are understood, then a potential approach would be to pre-compute these losses as a function of sensitive variables and store them for real-time lookup (e.g., tabular lookup). This will increase simulation fidelity with a negligible impact on execution speed.

1.3.11 Overly Simplistic Modeling of Radio Layers

It is a common practice for network simulations to not perform true bit-level simulations of the lower layers of the protocol stack. Rather, these lower layers are often abstracted into "clouds" with a static probability of performance metrics such as errors and delay. This approach is understandable given the challenges in bit-level simulations of large networks; however, this approach can lead to misleading results as it removes many dynamic aspects of system performance. It is important to understand the impact of these "averaging" approaches on simulation outputs and to manage expectations accordingly.

1.3.12   Disjoint M&S and Implementation Efforts

Too often M&S activities are disjoint from implementation efforts. This is unfortunate since a bit-true simulation can be a great interim milestone towards a real-world implementation and has the leave-behind value of a high-fidelity model. These activities should be tightly coupled. This is increasingly true as large companies continue to expand globally and development teams may be located on different continents instead of working side-by-side. While globalization has increased, so too have the tools to allow remote video teleconferences (VTCs) and information sharing. Hardware and software design tools such as LabVIEW Field Programmable Gate Array (FPGA) [134] or the Xilinx System Generator for Digital Signal Processing (DSP) Simulink blockset [133] also facilitate the conversion of a software model to a hardware implementation.

1.4   AN OVERVIEW OF COMMON M&S TOOLS

There are numerous network M&S tools available either as commercial products or as open source. This section provides a brief introduction to many of these tools. Table 1-3 provides a summary of many of the available network M&S tools [1].

Perhaps the four most commonly used network simulation tools in both academia and industry are Network Simulator 2 (NS-2), OPNET, QualNet, and GloMoSim. A short description of each follows.

1.4.1   NS-2

NS-2 is an open source DE simulator targeted at supporting network research. NS-2 is popular in academia because of its low cost (free) and extensibility. NS-2 was originally developed in 1989 as a variant of the REAL network simulator and, according to the NS-2 home project URL (see Table 1-3), "provides substantial support for simulation of TCP, routing, and multicast protocols over wired and wireless (local and satellite) networks."

NS-2 was built in the C++ programming language and provides a simulation interface through OTcl, an object-oriented extension of the scripting language Tcl. NS-2 will run on several forms of Unix (FreeBSD, Linux, SunOS, Solaris) and has been extended to Microsoft Windows (9x/2000/XP) using Cygwin (http://www.cygwin.com), which provides a Linux-like environment under Windows.

TABLE 1-3. Available Network Simulation Tools

Network Simulation Tool	URL
BRITE	http://www.cs.bu.edu/brite
Cnet	http://www.csse.uwa.edu.au/cnet/
GloMoSim	http://pcl.cs.ucla.edu/projects/glomosim
J-Sim	http://www.j-sim.org/
Matlab*	http://www.mathworks.com/products/matlab/
NS-2	http://www.isi.edu/nsnam/ns/
OMNeT++	http://www.omnetpp.org/
OPNET*	http://www.opnet.com/
PacketStorm-Network Emulator*	http://www.packetstorm.com/4xg.php
QualNet*	http://www.scalable-networks.com
Simulink*	http://www.mathworks.com/products/simulink/
SSFNet	http://www.ssfnet.org/homePage.html
x-sim	http://www.cs.arizona.edu/projects/xkernel/
NetSim*	http://www.tetcos.com/software.html
GTNetS	http://www.ece.gatech.edu/research/labs/maniacs/gtnets/index.html

*Denotes a commercial product.

NS-2 is currently licensed for use under version 2 of the GNU General Public License. Documentation has historically been poor for NS-2, with users left to rely on online user forums and newsgroups; however, there have been additional information sources emerging recently that may help someone new to NS-2, such as [6, 13].

1.4.2 OPNET

OPNET Technologies was founded in 1986, becoming a public company in 2000. The company provides a suite of software tools for network designers and administrators. But its flagship product is OPNET Modeler, which is a software tool for network M&S that was originally developed by the company's founder as a graduate project while at the Massachussetts Institute of Technology (MIT). OPNET Modeler is designed to either evaluate changes to existing networks or to design proprietary protocols. Furthermore, OPNET contains detailed models of specific network equipment. OPNET Modeler provides integrated analysis tools and a rich Graphical User Interface (GUI) as well as animation capabilities for data visualization. User development is in C/C++ and XML languages.

OPNET is slightly less common in academia as compared with NS-2, but is widely used in a variety of commercial and military organizations.

1.4.3 GloMoSim

The Global Mobile Information System Simulator (GloMoSim) is a DE simulator developed by the Parallel Computing Laboratory at UCLA in the C programming language and based on the parallel programming language Parsec. GloMoSim currently supports wireless protocols, which limits its utility in wired or hybrid networks. However, according to the GloMoSim project page (see URL in Table 1-3), there is currently development underway for a future revision that supports wired protocols. GloMoSim is available only to academic users; in fact, only users from an .edu domain are allowed to access the download page.

1.4.4 QualNet

QualNet is the commercial spin-off of the GloMoSim simulator offered by Scalable Network Technologies. QualNet is based on the C++ programming language and provides either command line or GUI interface to the user. QualNet provides a wide range of wired and wireless protocol support. Its key selling point is its high degree of scalability, which can supposedly "support simulation of thousands of network nodes" with high fidelity [16].

1.5 AN OVERVIEW OF THE REST OF THIS BOOK

This book takes a bottom-up approach to describing wireless network M&S, following the TCP/IP modified OSI stack model shown in Figure 1-4, recreated from [1].

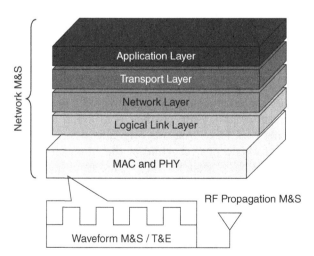

FIGURE 1-4. Wireless network simulation example demonstrating the interaction between various components [1].

This book first decomposes the wireless network M&S problem into a set of smaller scopes as depicted in Figure 1-4: 1) radio frequency (RF) propagation M&S (Chapter 2), 2) PHY M&S (Chapter 3), 3) MAC M&S (Chapter 4) and 4) higher layer M&S (Chapter 5). After considering each of these smaller scopes somewhat independently, the book then revisits the overall problem of how to conduct M&S of a wireless networking system in its entirety.

■■■■■■ CHAPTER 2

Modeling and Simulation for RF Propagation

All transmitted RF energy incurs path loss as electromagnetic waves propagate from source to destination. Propagation is a nontrivial problem because the exact path loss is completely dependent on a specific environment. A flat, desert environment has different propagation characteristics than a jungle environment, and a rural environment has different characteristics than a dense, urban environment. The goal of this chapter is not to make the simulation designer an expert in signal propagation, but to make the designer aware of the commonly used, underlying propagation models and choices for including these effects in a simulation. Network simulators such as NS-2, OPNET, GloMoSIM, and QualNET contain wireless package add-ons that allow large-scale PHY signal fading to be calculated for limited modeled scenarios. Although these capabilities are certainly better than neglecting the effects of signal fading and path loss all together, these models make many simplifying assumptions. Depending on the particular scenario being modeled, these assumptions may or may not be sufficient.

For example, consider the cellular engineer that desires to calculate a link budget between a base station (BTS) and mobile handset that is a line of sight (LOS) distance of 0.5 km from the BTS, where the BTS services a cell of radius 2 km. Clearly the handset is not at the edge of the cell and it would be expected that a link budget would contain a large error margin in this scenario. Here the cellular engineer does not require a high-fidelity answer and is probably not as concerned with intermittent small-scale fading effects such as Doppler or multipath as much as large-scale fading effects due to increasing transmitter-receiver distance. A Friis free space model or a two-ray model is sufficient to give the engineer a first-order approximation for the link budget. Now consider the same scenario but with a mobile handset located at the edge of a cell at approximately 2 km from the BTS. In this scenario, the engineer has little

An Introduction to Network Modeling and Simulation for the Practicing Engineer, First Edition. Jack Burbank, William Kasch, Jon Ward.
© 2011 Institute of Electrical and Electronics Engineers. Published 2011 by John Wiley & Sons, Inc.

flexibility built into his link budget and must be careful to consider all losses since such a scenario could potentially cause unnecessary handoffs that use precious resources of the base station controller (BSC) and core network. Both small-scale and large-scale fading must be considered in this scenario since all signal fluctuations could cause the received signal to interference noise ratio (SINR) to drop below the necessary threshold at the receiver.

The previous examples motivate the need for network designers to consider RF propagation models when applicable to the particular scenario being modeled. In power-limited wireless networks, especially those operating in the Industrial, Scientific, and Medical (ISM) bands and in heavy interference environments (i.e., low SINR), considering all fading mechanisms may be necessary; in other scenarios where sufficient SINR is available and achieving maximum throughput is not of primary concern, simplifying assumptions may have little impact on the end simulation results. When modeling RF propagation, more is not always better! Tools should be flexible enough to provide the simulation designer with knobs to turn such that the user can customize the experiment as necessary. However, too many customizable features can easily overwhelm the user. From the authors' experience, the most effective propagation simulators allow the user to begin with template scenarios and provide guidance for manipulating these example scenarios into the scenarios of interest.

The field of precise prediction of electromagnetic wave propagation easily deserves its own book devoted to this topic. In fact, [7] is one of the most widely cited resources on this topic and the interested reader is encouraged to study [7] for more details. This chapter is certainly applicable to wireless network simulation, but applies to the broader topic of simulating RF propagation environments. The intention of this chapter is to give the reader a clear understanding of the most popular underlying large-scale and small-scale RF propagation models and to present a variety of tools and corresponding capabilities. Specifically the wireless network simulation designer should develop a better understanding of the fading capabilities present in the popular network simulators and associated limitations. Indeed, all models are abstractions to the real world and the reader should expect RF propagation results to be scenario-dependent and based on a list of underlying assumptions. Also, the designer must not forget that between the particular device PHY layer being modeled and the air interface there is always an antenna with less-than-optimal characteristics. A simulation that considers all variables not only must take the antenna radiation pattern into account, but also the fact that many devices are not perfectly impedance matched.

This remainder of this chapter is organized as follows: the Fading Channel (Section 2.1), the ITU M.1225 Multipath Fading Profile for Mobile Wireless Interoperability for Microwave Access (WiMAX) (Section 2.2), Practical Fading Model Implementations—WiMAX Example (Section 2.3), RF Propagation Simulators (Section 2.4), and Propagation and Fading Simulations—Lessons Learned (Section 2.5).

2.1 THE FADING CHANNEL

When an electromagnetic wave propagates through a medium, it may experience reflection, diffraction, and scattering. Reflection occurs when an electromagnetic signal encounters an object such as the surface of the earth, buildings, or walls that have very large dimensions compared to the wavelength of the propagating wave. Diffraction occurs when the signal encounters an irregular surface with sharp edges that create a bending effect around the object. Scattering occurs when the medium through which the wave propagates contains a large number of objects smaller than the signal wavelength, such as foliage, street signs, and lamp posts [7, 8]. Propagation models generally fall into two categories: large-scale and small-scale models. Large-scale propagation models predict the mean signal strength for a given transmitter and receiver separation distance and are used to predict RF coverage. In the cellular example previously discussed, the Friis free space and two-ray models are large-scale propagation models that give the modeler an estimated path loss calculation under certain conditions. Small-scale propagation models characterize the rapid fluctuations of received signal strength over short distances or a short time duration. Small-scale models are generally associated with predicting multipath fading, or the effect of two or more copies of the transmitted signal combining at the receiver [7]. This section summarizes some of the most well-known and widely used large-scale and small-scale models in the research community, including analytical and stochastic models.

2.1.1 Large-Scale Fading

There are many large-scale propagation models that are used to calculate the path loss between a desired transmitter and receiver pair under various conditions. Before describing the models, there are a few path-loss conventions that should be described. First, path loss may be calculated as a positive or negative number, depending on how the equations are written and depending on how the results are used. A negative path loss is added to the total transmitter power to determine the received power level. A positive path loss is subtracted from the transmitter power to determine the received power level. Path loss and RF power in general is reported in units of decibels. It is assumed that the reader who is unfamiliar with dB notation can find a suitable tutorial and hence no overview of decibels is included here. Although Watts are typically used to report power, the Watt does not capture the large dynamic range of path loss and receiver sensitivity. As a rule, Wi-Fi and WiMAX radios have receiver sensitivities of approximately −85 dBm for the most robust modulation techniques (e.g., BPSK). Satellite Communications (SATCOM) systems can have receiver sensitivities in the range of −120 dBm and third generation (3G) cellular handsets typically receive signals of −75 to −80 dBm, although this depends on how the downlink (DL) power is measured in these Code Division Multiple Access (CDMA) systems. If we consider the case of a Wi-Fi or WiMAX system and assume that transmitted signals for these systems including the transmitter antenna gain are 1 Watt (i.e., 30 dBm), a path loss of 115 dB can be tolerated before the receiver sensitivity of −85 dBm is reached. The 115 dB of path loss is almost 12 orders of magnitude and would represent a number too large to practically handle in Watts (e.g., 316227766017 W). Additionally, multiplication and division operations in linear units such as Watts become additive and subtraction operations when working with decibels.

The most well-known propagation model is the Friis free space equation, shown below in Equation 2-1 [7]. Note that Equation 2-1 yields a positive path loss quantity and considers only the wavelength of the transmitted signal and the separation distance between the transmitter and receiver. The wavelength λ is computed by dividing the speed of light ($c = 3 \times 10^8$ m/s) by the transmit frequency f. There are many versions of the Friis free space equation that include the transmitter power (P_t), the transmitter antenna gain (G_t), and the receive antenna gain (G_r); however, the contributions of these components can be considered separately from the path loss, once the path loss has been computed as shown in Equation 2-2. Other well-known large-scale fading models are the two-ray ground reflection model and the lognormal shadowing model, both of which are described in sections 2.1.1.2 and 2.1.1.3, respectively.

2.1.1.1 *Free Space Path Loss*   The Friis Free Space Path Loss Model is described by Equation 2-1 [7].

$$FS(dB) = 10\log_{10}\left(\left(\frac{\lambda}{4\pi d}\right)^2\right) \tag{2-1}$$

where:

λ: The wavelength (m)
d: The transmitter to receiver separation distance (m)

$$P_r(dB) = P_t + G_t + G_r - FS(dB) \tag{2-2}$$

This model is used for simple path loss estimations because of its simple form and limited number of required parameters. The path loss exponent n is 2 in this case, denoting path loss that would only occur when LOS is available with no obstructions such as typically found with SATCOM and certain microwave point-to-point links. The application of the free-space path loss equation to wireless network scenarios is questionable given that most of these systems operate in environments with many RF obstacles. Consider modeling a WiMAX network in an urban setting where a single base station services 100 fixed subscribers. In this case, the path loss between the base station and a given subscriber would not be expected to follow the Friis free space equation.

The path loss exponent can be increased to represent environments other than the LOS free space environment, but the Friis free space equation still only considers a single transmission path between transmitter and receiver. That is, the Friis equation does not consider multipath, just the mean path loss over a distance at a given frequency. Table 2-1 lists common path loss exponents that researchers have applied to different environments.

2.1.1.2 Two-Ray Ground Reflection Model The two-ray model is a commonly used propagation model because it accounts for a ground-reflected path between transmitter and receiver in addition to the LOS component. The two-ray model has been shown to produce more accurate path loss estimates at long distances than the Friis free space equation. Moreover, the two-ray model also accounts for antenna height differences at the transmitter and receiver, which is not considered in the Friis equation [7]. Figure 2-1 illustrates the geometry of the two-ray ground reflection model applied to an example transmitter and receiver separation [115].

The two-ray ground reflection model is most often used in the form shown in Equation 2-3 [7].

$$PL(dB) = 40\log_{10}(d) - (10\log_{10}(G_t) + 10\log_{10}(G_r) + 20\log_{10}(h_t) + 20\log_{10}(h_r))$$
$$(2\text{-}3)$$

TABLE 2-1. Path Loss Exponents for Different Environments [7]

Environment	Path Loss Exponent, n
Free space	2
Urban area cellular radio	2.7 to 3.5
Shadowed urban cellular radio	3 to 5
In building line-of-sight	1.6 to 1.8
Obstructed in building	4 to 6
Obstructed in factories	2 to 3

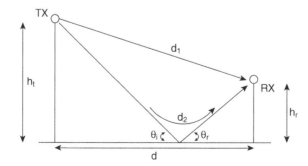

FIGURE 2-1. Two-ray ground reflection model illustration recreated from [115].

where,

d: The transmitter to receiver separation distance (m)
G_t: The transmitter antenna gain
G_r: The receiver antenna gain
h_t: The transmit antenna height (m)
h_r: The receive antenna height (m)

Note that Equation 2-3 is independent of frequency and only holds true at large distances defined as $d \gg \sqrt{h_t h_r}$. At small transmitter and receiver separation distances, a series of electromagnetic field equations must be used to calculate the total E-field and account for constructive and destructive interference of the E-field that occurs at short distances. A discussion of these equations is beyond the scope of this chapter and the interested reader is encouraged to read [7].

2.1.1.3 Log-Distance Path Loss Model with Shadowing   In both indoor and outdoor channels, theoretical and measurement-based propagation models indicate that average received power decreases logarithmically with distance. Measurement campaigns have shown that because different environments have different obstructions between the transmitter and receiver, a lognormally distributed random variable can be used to characterize the shadowing effects that occur with mean value determined by the transmitter and receiver separation distance [7]. Equation 2-4 presents the lognormal shadowing equation.

$$L(dB) = FS(d_0) + 10n\log_{10}\left(\frac{d}{d_0}\right) + X_\sigma \qquad (2\text{-}4)$$

where,

d_0:   is the close-in reference distance and is assumed to be 100 m (to avoid near-field effects)
FS: is a free space path loss calculated from Equation 2-1 at distance d_0
n: the path loss exponent, unique for each radio environment to be modeled
d: the transmitter to receiver separation distance in km
X_σ: is a zero-mean Gaussian distributed random variable (in dB) with standard deviation σ (also in dB).

The lognormal distribution means that in units of dB, X_σ follows a Gaussian distribution, with probability distribution function (PDF) specified by Equation 2-5.

$$p(x) = \frac{1}{\sqrt{2\pi\sigma^2}}e^{-\frac{(x-m)^2}{2\sigma^2}} \qquad (2\text{-}5)$$

where m is the mean.

Note that the mean m and standard deviation σ are both specified in units of dB. The simulation designer must choose a path loss exponent n from Table 2-1 and standard deviation σ that apply to the simulation scenario being modeled. X_σ is site and distance dependent with typical values ranging from 6 to 10 dB for urban environments [7].

2.1.1.4 Wireless Network Simulators with Large-Scale Fading Models

The previous three sections present three large-scale fading models commonly used in both analytical estimates and simulation. The reader most likely has two questions. First, how important is the inclusion of large-scale fading on the simulated output? And second, how do the three models presented help the simulation designer incorporate large-scale fading into his or her simulation? The first question is best answered with an example from [9], where the authors compare experimental data to common simulation assumptions. One such common simulation assumption is that if the receiver receives the transmitted signal at all, it is received without error. That is, large-scale fading is not considered at the receiver, only the binary function that within a threshold distance, the desired signal is received with probability 1 and outside of that threshold the signal is received with probability zero. The authors of [9] collected experimental data from IEEE 802.11 broadcast beacons at various transmitter to receiver separation distances. Figure 2-2, recreated from [9],

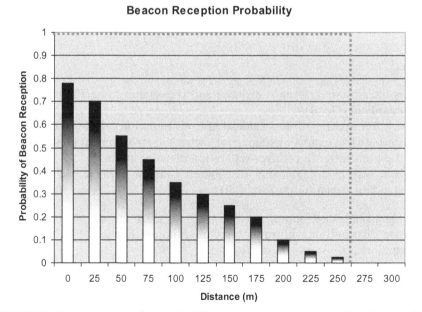

FIGURE 2-2. Beacon reception probability versus transmitter to receiver distance [9].

illustrates the observed probability of beacon reception as the distance between transmitter and receiver increases. As would be expected with any of the three large-scale propagation models presented in this section, the signal strength decreases exponentially as the distance increases and clearly does not adhere to the often-assumed binary function outlined in yellow in Figure 2-2. Recall, however, that the degree to which the reception probability decays is a function of the path loss exponent, which is dependent on the environment being modeled. Furthermore, these measurements also account for small-scale fading effects as discussed in the next section.

The previous example motivates the need to consider large-scale fading effects in simulations, but how does the simulation designer incorporate these into his or her model? This depends on the simulation designer's objectives. The three large-scale fading models described are relatively straightforward to implement in a homebrew simulation such as a Matlab, C, or Java simulation. In this case the designer must consider the particular scenario being modeled. Do all of the wireless nodes have LOS with all other nodes? If this is the case, then the Friis free space equation or two-ray model may be sufficient. Is there a significant difference in the antenna heights between transmitting nodes? This difference is not captured in the Friis equations, but is a parameter in the two-ray model. In the case where LOS does not hold and there are path obstructions between nodes, the lognormal shadowing model is probably best suited. The geometry of the scenario being modeled is driven by acceptable separation distance between transmitter and receiver. Ultimately, this is driven by the receiver sensitivity of the radios being modeled and the maximum path loss that can be incurred. Because antennas are not considered in the Friis equation and more generally because of its simplistic form, it may produce an overly optimistic path loss for the desired user. Although this may produce an advantage in terms of received power for the desired transmitter, it is not necessarily an increase in SINR; interferers that also experience a favorable free-space path loss can raise the noise floor at the desired receiver by arriving more powerfully than expected [10].

The simulation designer must consider the acceptable margin of error in his or her simulation. The Friis free space path loss model is simplistic, but if a large amount of margin is acceptable in the accuracy of the prediction, then the time saved implementing the Friis equation versus other more detailed models may be worthwhile. Alternatively, if important decisions such as the purchase of amplifiers and antennas are dependent on the prediction of simulation results, then a more detailed simulation model should be considered. The lognormal shadowing model is used by many designers as a compromise between implementation time and detail. This model considers path obstructions and includes contributions from random fading, but is relatively simplistic in its implementation. The generation of Gaussian and lognormal random variates is well known and available in sources such as [11, 12]. Additionally, [12] includes C and C++ programs for random variate generation. The designer must also consider which models are available in his

TABLE 2-2. Large-Scale Propagation Models Available in GloMoSim, NS-2, OPNET, and QualNET Recreated from [10]

GloMoSim (v.2.03)	NS-2 (v. 2.1b8)	OPNET	QualNET
Free space, two ray [14]	Free space, two ray, lognormal shadowing, Nakagami [13]	Free space, two ray*, TIREM, Longley-Rice, CCIR, Hata, COST 231 Walfisch-Ilegami [15]	Free space, two ray, TIREM, COST 231-Hata, COST 231-Wi, Longley-Rice [16]

*Denotes this model and support as being provided by user community.

or her favorite simulator. The spin-up time required to become familiar with a new simulator may not be worth the extra detail provided by additional supported fading models.

The four simulators referenced throughout this book offer support for the free space path loss model, two-ray model, and NS-2 supports lognormal shadowing as shown in Table 2-2. NS-2 also supports a distance threshold model such as the one described in the previous example, that the authors discourage the reader from using [13].

For further information on the specific details of incorporating the large-scale fading models into a wireless network simulation, see [14] for GloMoSim and [13] for NS-2 information. The authors are unaware of publicly available OPNET and QualNET manuals since these are commercial products. The QualNET library contains the following additional large-scale fading models: Terrain Integrated Rough Earth Model (TIREM), European Cooperation in Scientific and Technical Research (COST) 231-Hata, COST 231-Wi, and Longley-Rice [16]. The OPNET library also includes the following large-scale fading models: CCIR, Free Space, Hata, Longley-Rice, TIREM v3 and v4, and Walfisch-Ilegami [15]. The respective companies should be contacted for additional information and technical support.

2.1.2 Small-Scale Fading

Small-scale propagation models characterize the rapid fluctuations of received signal amplitudes, phases, or multipath delays over short distances or a short time duration. Large-scale fading accounted for path loss between the transmitter and receiver and models provide a general estimate of the mean signal strength available at the receiver. Small-scale fading accounts for significant degradation at the receiver since quick fluctuations in signal strength can corrupt bits in digital communication systems and therefore cause degradation to higher-layer protocols in a wireless network. Multipath interference, caused by the interaction of multiple copies of the transmitted signal arriving at the

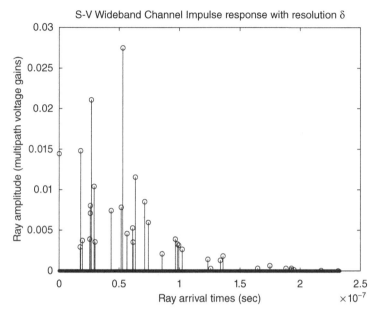

FIGURE 2-3. Example simulated instantaneous power delay profile using the Saleh-Valenzuela wideband channel model [114].

receiver via different paths, can add constructively or destructively, hence improving or degrading signal reception. Mobility can also affect small-scale fading, where Doppler shifts can induce random frequency modulation into the desired signal [7]. As with large-scale fading, there are many commonly used small-scale fading models in the simulation community. This section describes the most popular models, including the Ricean and Rayleigh distributions.

There are many metrics that describe the characteristics of various fading channels. An in-depth treatment of all topics concerning fading channels is beyond the scope of this book and references such as [7] should be consulted if more details are desired; however, some details describing the time-dispersive fading channel characteristics are included here to equip the reader with the background to potentially consider these characteristics in future simulations and to better explain the mapping between simulation results and real-world experimentation. The power delay profile characterizes the instantaneous arrival of multipath components at the desired receiver from the first arrival to the last arrival, where the y-axis demonstrates the relative power of each multipath component. Figure 2-3 illustrates a simulated example power delay profile.

Various parameters such as the mean excess delay and root mean square (RMS) delay spread may be generated from the power delay profile to characterize the channel completely. The mean excess delay $\overline{\tau_i} = \sum_k a_k^2 \tau_k / \sum_k a_k^2$

captures the mean time that significant multipath components will arrive at the desired receiver after the first component is received. The RMS delay spread or the standard deviation of the multipath energy is expressed in Equation 2-6 as

$$\sigma_{\bar{\tau}_t} = \sqrt{\overline{\tau_t^2} - (\overline{\tau}_t)^2}, \tag{2-6}$$

where $\overline{\tau_t^2} = \sum_k a_k^2 \tau_k^2 / \sum_k a_k^2$. In all equations, a_k represents the amplitude of the tapped-delay line channel model multipath component with delay τ_k. A temporal or spatial average power delay profile may also be calculated from many instantaneous power delay profiles.

The simulation designer is not expected to generate these types of statistics for a given scenario, but should attempt to match any channel delay parameters as accurately as possible to the environment being simulated. The RMS delay spread is used to calculate a statistic called the coherence bandwidth, which is a defined relationship inversely proportional to the RMS delay spread of a channel. The coherence bandwidth indicates the range of frequencies over which the channel can be considered flat or the channel passes spectral components with approximately equal gain and linear phase [7]. In many wireless network simulations, the designer is concerned with generating higher layer statistics and uses standards-based PHY waveforms such as the 22 MHz IEEE 802.11b waveform or the variable-sized IEEE 802.16-2004 waveforms. This is popular practice and presents no problems as long as the designer is aware that to prevent distortion in a channel, the particular transmitted waveform must not exceed the coherence bandwidth of the channel. This becomes increasingly important as data rates and improvements in pulse-generating hardware decrease the duration of pulses and therefore approach the threshold of true ultra wideband (UWB) signals that contain components that occupy bandwidths on the order of 1 GHz. For these systems, flat-fading simulations may not be accurate because of the signal's large baseband bandwidth compared to the coherence bandwidth of the channel.

Frequently wireless network simulation designers make simplifying assumptions on the channel conditions such as the duration time of a packet is less than or equal to the coherence time of the channel. Doppler spread and coherence time describe the time-varying nature of a signal experiencing small-scale fading, just as the delay spread and coherence bandwidth describe the time-dispersive nature of the channel in a local area [7]. The Doppler shift is a frequency change that occurs when an electromagnetic wave is received by an object in motion as illustrated in Figure 2-4. The Doppler shift is calculated in Equation 2-7. Note that a signal received that is orthogonal to the direction of velocity produces no Doppler shift.

$$B_D = f_D = \frac{v}{\lambda} \cos(\theta) \tag{2-7}$$

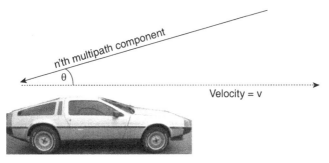

FIGURE 2-4. Example of Doppler shift.

TABLE 2-3. Summary of IEEE 802.11b and g PHY Parameters and Common 2.4 GHz Indoor Channel Parameters [104]

2.4 GHz Channel Parameters	Indoor Environment	Signal Parameters	IEEE 802.11b and g
Coherence time (ms)	21.2–212	Minimum symbol time (ns)	b) 90 g) 888
Coherence BW (MHz)	3–55	Bandwidth (MHz)	b) 22 g) 16.250
RMS delay spread (ns)	45–420	Subcarrier spacing (kHz)	b) Not OFDM g) 312.5
Mean excess delay (ns)	12–42		

The coherence time of the channel is then inversely related to Doppler spread B_D and captures the time duration over which the channel impulse response is invariant and hence two received signals should have strong amplitude correlation. The channel will remain relatively constant over the transmission of a baseband message if the inverse of the baseband signal bandwidth to be transmitted is less than the coherence time of the channel. For example, if an IEEE 802.11b signal that occupies 11 MHz of baseband bandwidth is to be transmitted and the channel is to be assumed invariant, the coherence time of the channel must be greater than 1/11 MHz or approximately 90 ns. Table 2-3 summarizes a set of channel characterization parameters derived from a 2.4 GHz measurement campaign described in [104]. These parameters describe one possible time-varying channel and should be interpreted as one set of possible indoor channel parameters.

The simulation designer may be given similar parameters and asked to make performance predictions for a wireless network in a similar environment. The intention is that the comparison in Table 2-3 should help the simulation designer better understand the limitations of his or her simulations and possible areas for troubleshooting, but in the most general case the resulting effects of delay spread and mobility are captured in small-scale fading models such as Ricean and Raleigh already available in most simulators. Continuing with the IEEE 802.11 example, for the mobile indoor environments considered in [104], the minimum observed coherence time exceeded the minimum IEEE 802.11b symbol time of 1/11 MHz. This means that the IEEE 802.11b signal will experience slow fading and the channel will remain relatively constant over the transmission of the message. From Table 2-3, the IEEE 802.11 b and g channel bandwidth both exceeds and is below the range of the fading channel bandwidth (coherence bandwidth). This means that the IEEE 802.11 signal will experience a combination of flat and frequency-selective fading. The IEEE 802.11b signal falls within the range of the channel delay spread in Table 2-3, which indicates a combination of flat and frequency selective fading; the IEEE 802.11g signal exceeds the channel delay spread because of its relatively slow symbol time, which is a main factor driving the implementation of orthogonal frequency division multiplexed (OFDM) signals in wireless networking devices (e.g., IEEE 802.11a,g,n, IEEE 802.16, 4G cellular Long Term Evolution (LTE) advanced).

As a final overview to the different types of small-scale fading experienced by a transmitted signal, Figure 2-5 summarizes the characteristics of the two types of fading due to multipath time delay spread, flat fading and frequency selective fading, and the two types of fading due to Doppler spread, fast fading and slow fading. The motivation for including Figure 2-5 is twofold. First, this summarizes the fact that small-scale fading takes many forms and must be modeled carefully if the simulated small-scale fading environment is to match the actual environment of interest. The second motivation for including Figure 2-5 is to give the simulation designer a roadmap that summarizes the associated tradeoffs. The ability to effectively troubleshoot a simulation is as important and, in some cases, more time-consuming than the original simulation implementation. Figure 2-5 provides a starting place for the simulation designer to begin troubleshooting his or her simulation if the effects of small-scale fading are in question by providing a sanity check mechanism to map the simulation to a real-world environment, as introduced by the example describing Table 2-3.

In flat fading channels, all frequency components that compose a signal undergo the same fading, yet the signal strength of the received signal changes over time due to multipath. Frequency selective channels affect a signal's frequency components differently and the signal becomes distorted at the receiver due to intersymbol interference (ISI). Most simulations assume a flat fading channel since frequency selective channels are difficult to model [7, 116]. In a fast fading channel, the channel impulse response simply changes rapidly within one symbol duration of the transmitted signal (i.e., the coherence time of the channel is smaller than the symbol period of the transmitted signal). Conversely, in a slow fading channel, the channel impulse response changes slowly compared with the transmitted signal. The four types of small-

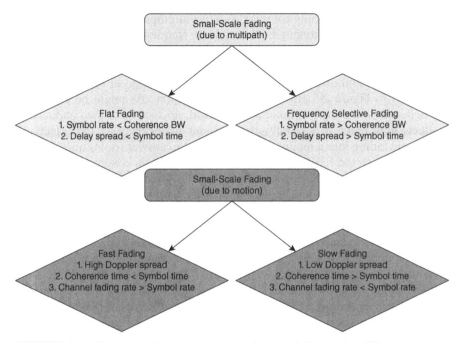

FIGURE 2-5. Flow chart demonstrating the characteristics of the different types of small scale fading. Recreated from [116].

scale fading are pairwise mutually exclusive, but a channel can be described by both multipath effects due to the delay spread and the Doppler spread of the channel. That is, flat fading and frequency selective fading are mutually exclusive as well as fast and slow fading; however, a channel may be flat and fast fading or frequency selective and fast fading depending on the particular channel characteristics [7, 116].

Figure 2-5 contains the list of questions to be asked about the real-world fading environment before blindly implementing the default fading model in a simulator. For example, if the simulation designer is told to model a stationary IEEE 802.11b WLAN in an urban environment, a good assumption would be to model a slow, frequency selective fading environment. The stationary network in most cases makes the coherence time of the channel greater than an IEEE 802.11b symbol period, yet the relatively large bandwidth of an IEEE 802.11b waveform likely exceeds the coherence bandwidth of the channel in an urban environment. This is of course assuming that the inverse delay spread of the channel is greater than a symbol period. There are general rules of thumb for various networking systems that can provide guidance on the conditions of a given propagation environment; this is especially true for the cellular industry, which has not only compiled significant empirical data that describe the propagation environment in the cellular frequency bands, but has RF propagation software packages such as those described at the end of this chapter.

The next natural question is how does the information in this section and specifically in Figure 2-5 affect an actual simulation? As might be expected, this depends on the exact simulator being used, since some of the commercial products may make it easier for the designer to incorporate the effects of small-scale fading into a model. In formulating the answer to this question, let us continue with the IEEE 802.11 example and make the assumption that the wireless channel is in fact a slow, frequency selective fading channel. From Figure 2-5 and the previous discussion, the bandwidth of the simulated signal exceeds the coherence bandwidth of the channel. There is not much for the simulation designer to alter in terms for the transmitted signal. An IEEE 802.11 direct sequence spread spectrum (DSSS) signal occupies a bandwidth of 22 MHz per the IEEE 802.11 standard [17]. The data rate of an IEEE 802.11 signal is also specified in [17], so the designer cannot simply modify the data rate to change the symbol period. But the designer has control over creating the multipath environment through the Ricean and Rayleigh fading distributions discussed in the next sections. In this example, the designer must generate the fading environment such that the delay spread of the channel exceeds the IEEE 802.11 symbol period for the data rate under simulation. Because the network is stationary, the coherence time of the channel is assumed larger than the period of a symbol. If a fast fading channel is needed for a particular scenario, the mobility options of the designer's preferred simulator must be incorporated.

2.1.2.1 The Ricean Fading Distribution Unlike large-scale fading models, small-scale fading models are stochastic and based on the work of previous researchers that matched empirical data from various measurement campaigns to statistical random variable distributions to find curves that best fit. The Ricean distribution is widely accepted by researchers as representative of fading scenarios where a LOS component is present in the signal of interest. In this case, the random multipath components are superimposed on a dominant LOS component. The Ricean distribution is commonly described in terms of a parameter K that is often referred to as the "Ricean K factor," shown in Equation 2-8 [7, 18].

$$K(dB) = 10\log_{10}\left(\frac{A^2}{2\sigma^2}\right) \qquad (2\text{-}8)$$

where,

A: the amplitude of the LOS component present in the signal of interest
σ^2:   the variance of the multipath interference present at the desired receiver

Equation 2-9 presents the PDF for a Ricean distributed random variable [7, 18].

$$p_{Ricean}(r) = \begin{cases} \dfrac{r}{\sigma^2} e^{-\frac{(r^2+A^2)}{2\sigma^2}} I_0\left(\dfrac{A_r}{\sigma^2}\right) & for\ (A \geq 0, r \geq 0) \\ 0 & for\ (r < 0) \end{cases} \qquad (2\text{-}9)$$

where,

r: the received envelope voltage of a multipath component
I_0:   the modified Bessel function of the first kind and zero order

Figure 2-6 illustrates the Ricean PDF for three different K factors, K = 0 dB, K = −55 dB, and K = −∞. Note that from Equation 2-9, when the LOS component fades below the level of the non LOS (NLOS) multipath interference

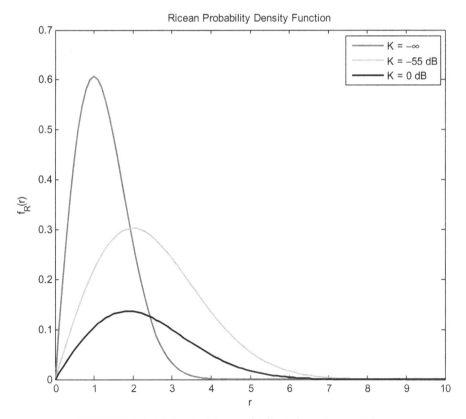

FIGURE 2-6. PDF of a Ricean distributed random variable.

(i.e., A = 0), the Ricean distribution degenerates to the Rayleigh distribution (discussed later), which is the case where K = −∞.

2.1.2.2 The Rayleigh Fading Distribution In wireless communications systems, the Rayleigh distribution is often used to model multipath interference when no dominating LOS component is present. That is, all signal arrivals at the receiver come from NLOS paths. The justification for modeling the multipath interference as Rayleigh comes from the fact that the sum of two quadrature Gaussian noise signals follow a Rayleigh distribution [106]. As previously noted, the Rayleigh distribution is a special case of the Ricean distribution, when the K factor amplitude A is set to zero. Equation 2-10 presents the PDF for a Rayleigh distributed random variable.

$$p_{Rayleigh}(r) = \begin{cases} \dfrac{r}{\sigma^2} e^{-\frac{r^2}{2\sigma^2}} & (0 \le r \le \infty) \\ 0 & (r < 0) \end{cases} \tag{2-10}$$

2.1.2.3 Wireless Network Simulators with Small-Scale Fading Models
At this point the reader should be familiar with a few key concepts concerning signal propagation and how it relates to the simulation of wireless networks. RF signals undergo significant fading, both large scale and small scale, that affects a receiver's ability to correctly demodulate and decode a frame's contents. Multipath interference should not be confused with interference as described in Chapter 3. Multipath interference is certainly related to the broader topic of interference in an environment where multiple users are generating traffic, yet multipath components are also generated by only the desired signal in a single-user scenario. This is an important concept to understand because even in the simplistic case of a single ideal user, the desired user's signal is still degraded by large-scale and small-scale fading. But the question remains exactly how do small-scale fading models apply directly to simulation? This depends on whether the reader intends to create his or her own homebrew simulation including fading or if the reader intends to use one of the well-known network simulators previously discussed.

The Ricean and Rayleigh fading distributions represent the random signal amplitudes experienced at a receiver in a given environment due to fading. Because these small-scale models are random variable distributions, the various multipath components must be modeled as random variables. In the most general case, this is accomplished by generating random variates that follow the desired distribution (e.g., Ricean or Rayleigh). A uniformly distributed pseudo random number generator that generates numbers on the interval of [0,1] can be used to generate samples of another distribution through the inverse cumulative distribution function (CDF) of the desired distribution [11]. These inverse CDFs are well-known for common probability distributions and available in tables such as the one in [11]. For example, Equation 2-11 shows the inverse CDF to convert uniformly distributed random variates on the interval of [0,1] to Rayleigh random variates.

$$X = F^{-1}(U) = \sigma\sqrt{-\log(1-U)} = \sigma\sqrt{-\log(U)} \qquad (2\text{-}11)$$

where,

U: the uniformly distributed random variable on the interval of $[0, 1]$
σ: the standard deviation of the multipath interference present at the desired receiver

Furthermore, Equation 2-11 is simplified noting that 1—U has the same uniform distribution as U, on the interval of [0,1]. In a practical simulation application, this means that the simulation designer can use this equation to convert the output of the pseudo random number generator that is between 0 and 1 to Rayleigh distributed samples that can be added directly to the desired and interferer signals.

Of course, it is not quite so simple because there is one missing component that has a huge impact on the output. Because the uniformly distributed random variates are uncorrelated, the resulting random variates, Rayleigh from the example, are also uncorrelated. In a real-world scenario, multipath components are naturally correlated because multiple reflections may occur due to the same obstacle. There is, in fact, a time correlation of the signal envelope that is not captured when the generated random variates are uncorrelated. In a wireless network simulation, the time correlation of the signal envelope could account for burst errors across a frame that may otherwise be simulated as random errors if the multipath samples are uncorrelated [18]. There are approaches to force the correlation of random variates described in references such as [11], but these are beyond the scope of this chapter and generally have significant computational requirements that may not scale in a large wireless network simulation.

The approach used by the common network simulators and described in [18] is to pre-compute a data set of multipath components that are time correlated based on time-averaged power, maximum Doppler frequency, and Ricean K factor. This sequence is constructed such that it can be repeated without discontinuities and large simulations may repeat the sequence multiple times as necessary. The statistical validity of this approach is demonstrated in [18]. The effects of large-scale fading are also incorporated into this model as affecting the mean received power and hence the mean of the Ricean or Rayleigh distributed multipath components. The designer is cautioned to carefully index the table lookup if he or she chooses to implement small-scale fading using the methods described in [18]. In discrete event simulation, if indexes are chosen based on time instant alone, the desired signal and all interferer signals may experience the same fading in the simulation. This is not representative of a real-world scenario and should be avoided. The desired

TABLE 2-4. Small-Scale Fading Models Available in GloMoSim, NS-2, OPNET, and QualNET. Recreated from [10]

GloMoSim (v.2.03)	NS-2 (v. 2.1b8)	OPNET	QualNET
Ricean, Rayleigh [10, 14]	Ricean, Rayleigh [13, 18]	Ricean, Rayleigh* [15]	Ricean, Rayleigh [16]

*Denotes this model and support as being provided by user community

signal and interferers should be labeled and handled separately in the simulation such that table lookups into the Ricean fading statistics can use separate indexes for separate users. This will allow all users in the simulation to remain uncorrelated and produce more valid results.

For the simulation designer that desires to choose one of the small-scale fading models available within one of the four well-known network simulators, Table 2-4 lists the small-scale fading models present in specific versions of GloMoSim, NS-2, OPNET, and QualNET. As can be seen, all support Ricean and Rayleigh fading, implemented as described in the previous paragraph and in detail in [18].

2.2 THE ITU M.1225 MULTIPATH FADING PROFILE FOR MOBILE WiMAX

This section introduces the reader to a frequently encountered, empirically determined channel model referred to as the International Telecommunication Union (ITU) M.1225 and defined in [93]. The ITU M.1225 channel models are used by the WiMAX Forum in the conformance testing process to determine interoperability of Mobile WiMAX equipment in fading environments [96]. Because the WiMAX (IEEE 802.16) technology is used as an example in subsequent chapters of this book to describe different aspects of wireless network M&S, the ITU M.1225 model in the context of Mobile WiMAX is described. The ITU M.1225 document itself is stand-alone and radio transmission technology (RTT) agnostic, setting forth in generic terms the recommended channel models (propagation path loss equation for large-scale fading component and channel impulse response for small-scale fading component) for evaluating terrestrial and satellite communications conforming to the ITU's IMT-2000 3G specifications. Only the terrestrial channel models are considered in this chapter. As previously described, channel models can only capture the statistical characteristics of a channel and do not exhaustively represent all possible operating environments for a cellular system. The intention of the ITU M.1225 document is that the example channel models sufficiently represent a set of worst-case operating environments such that mobile equipment operating correctly in this environment would therefore also operate correctly in less stressful environments. The ITU M.1225 channel models characterize three different operating environments: 1) an indoor office environment, 2) an outdoor to indoor and pedestrian environment, and 3) a vehicular environment. Of the three test environments, Mobile WiMAX compliance requires only the outdoor to indoor and pedestrian environment (pedestrian environment) and the vehicular environment.

2.2.1 ITU M.1225 Large-Scale Path Loss Models

The large-scale path loss for the ITU M.1225 pedestrian environment and vehicular environment are described by two equations generated through curve-fitting methods applied to measured data collected in the 2 GHz frequency band. Both equations are presented here for completeness [93].

Path Loss Model for Pedestrian Environment:

$$L(dB) = 40\log_{10} R + 30\log_{10}(f) + 49 \tag{2-12}$$

where,

R: BTS to mobile station (MS) separation distance (in km)
f: Carrier frequency in MHz of system being simulated (should not deviate far from 2 GHz)

Path Loss Model for Vehicular Environment:

$$L(dB) = 40(1 - 4 \times 10^{-3}\Delta h_b)\log_{10} R - 18\log_{10}(\Delta h_b) + 21\log_{10}(f) + 80 \tag{2-13}$$

where,

R: BTS to MS separation distance (in km)
f: Carrier frequency in MHz of system being simulated (should not deviate far from 2 GHz)
Δh_b: BTS antenna height (m), measured from the average rooftop level (15 m is general recommendation, valid for $0 \le \Delta h_b \le 50$ m)

The simulation designer that desires to incorporate these large-scale path loss models into his or her simulation should be familiar with the model limitations. Fundamentally, these models are derived from measurement campaigns that utilized approximately a 2-GHz carrier frequency. This means, as noted in the description for f, the model should not be applied to frequencies that deviate too far away from 2 GHz. How much deviation is too far? As a general rule, the authors recommend not applying this model to frequencies below 1.5 GHz or above 2.5 GHz without measurement-based verification of the model's validity at those frequencies. Additionally, care should be taken with regard to adjusting the antenna height parameter Δh_b in the vehicular model to match the scenario being simulated; that is, Δh_b occurs twice in Equation 2-13, both inside and outside of a logarithmic function. An error in Δh_b by ±5 m, results in at least a 3 dB change in the resulting path loss $L(dB)$. The pedestrian model assumes a worst-case NLOS path and lognormal shadowing with a standard deviation of 10 dB for outdoor users and 12 dB for indoor users. The vehicular model assumes a worst-case NLOS path and lognormal shadowing with a standard deviation of 10 dB for both an urban and suburban area. The interested reader is referred to [93] for more details.

2.2.2 ITU M.1225 Small-Scale Path Loss Models

The small-scale fading component of the ITU M.1225 pedestrian environment and vehicular environment channel models are described by the parameters in Tables 2-5 and 2-6, respectively. Each channel model is characterized by a channel impulse response modeled as a tapped-delay line, specified by Equation 2-14.

$$h(t) = \sum_{k=0}^{L-1} a_k \delta(t - k\Delta\tau) \tag{2-14}$$

where,

a_k: amplitude of kth resolvable multipath component

δ: unit impulse marking the arrival of a multipath component at relative delay τ with bin duration $\Delta\tau$

L: Maximum relative delay of resolvable multipath components for a given impulse response iteration

The relative spacing of the taps (relative delay), average power relative to the strongest tap (average power (dB)), and the Doppler spectrum of each tap are provided in Tables 2-5 and 2-6 [93]. Two impulse responses are specified for the pedestrian and vehicular channel models that represent a small RMS delay spread (channel A) and a median RMS delay spread (channel B); the RMS delay spread can be computed from the channel impulse response information in Tables 2-5 and 2-6 using Equation 2-6 and is provided in [93].

In the WiMAX forum conformance test description, only channel B of the pedestrian model (Pedestrian-B) with a velocity of 3 km/hr and channel A of the vehicular model (Vehicular-A) with a velocity of 60 km/hr and 120 km/hr are used [96].

The specified velocities introduce the effects of mobility into the model by defining the Doppler spectrum. Equation 2-15 approximates the Doppler

TABLE 2-5. Small-Scale Fading Parameters for ITU M.1225 Pedestrian Environment Tapped-Delay Line

Tap	Channel A		Channel B		Doppler Spectrum
	Relative Delay (ns)	Average Power (dB)	Relative Delay (ns)	Average Power (dB)	
1	0	0	0	0	Classic
2	110	−9.7	200	−0.9	Classic
3	109	−19.2	800	−4.9	Classic
4	410	−22.8	1200	−8.0	Classic
5	—	—	2300	−7.8	Classic
6	—	—	3700	−23.9	Classic

TABLE 2-6. Small-Scale Fading Parameters for ITU M.1225 Vehicular Environment Tapped-Delay Line

Tap	Channel A		Channel B		Doppler Spectrum
	Relative Delay (ns)	Average Power (dB)	Relative Delay (ns)	Average Power (dB)	
1	0	0	0	−2.5	Classic
2	310	−1.0	300	0	Classic
3	710	−9.0	8900	−12.8	Classic
4	1090	−10.0	12900	−10.0	Classic
5	1730	−15.0	17100	−25.2	Classic
6	2510	−20.0	20000	−16.0	Classic

spectra for both the Pedestrian and Vehicular environments, where the Doppler shift f_D is determined from Equation 2-7 using the classic (Jakes) spectrum.

$$S(f) \approx \frac{1}{\sqrt{\left(1 - \left(\dfrac{f}{f_D}\right)^2\right)}} \quad \text{for} \quad f \in [-f_D, f_D] \qquad (2\text{-}15)$$

The Doppler spectrum indicates the frequency deviation away from the carrier frequency that should be expected and tolerated by WiMAX equipment in a mobile scenario. The classic Doppler spectrum is not the only choice; other Doppler spectrum types are defined such as Gaussian, flat, and rounded. The particular scenario being modeled shall determine the correct Doppler spectrum choice. In many cases, this parameter is specified in a well-known model such as the ITU M.1225, or a source such as [7] could be used to match an appropriate Doppler spectrum to the environment of interest. The simulation

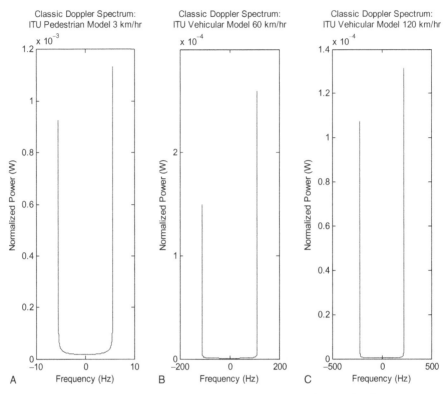

FIGURE 2-7. Classic Doppler spectra for A) Pedestrian-B (3 km/hr), B) Vehicular-A (60 km/hr), C) Vehicular-A (120 km/hr).

designer wants the Doppler spectrum to be as tight around the carrier frequency as possible for a given mobility scenario since this allows more flexibility for maintaining synchronization on the DL; however, the simulation designer must also be careful to consider the possible worst-case f_D for the maximum specified operational velocity to avoid scenarios where mobile systems are unable to maintain synchronization and unexpected failures occur. Figure 2-7 illustrates the classic Doppler spectra for the three WiMAX conformance test environments [96].

2.3 PRACTICAL FADING MODEL IMPLEMENTATIONS— WiMAX EXAMPLE

Undoubtedly by this point the reader has been sufficiently motivated by the previous sections as to the importance of considering fading models when developing a wireless network simulation. In many cases, the inclusion in a

TABLE 2-7. Mobile WiMAX channel models available in GloMoSim, NS-2, OPNET, and QualNET

GloMoSim	NS-2 (v. 2.31)	OPNET	QualNET
No known WiMAX channel models	WiMAX Model version 2.6 [99]: COST 231, ITU M.1225 Pedestrian and Vehicular Models [95, 105]	WiMAX (802.16) Specialized Model for OPNET Modeler Wireless Suite: ITU M.1225 Pedestrian and Vehicular Models [98]	Advanced Wireless Library: COST 231 Walfish-Ikegami, Street Microcell Models [97]

given design of basic fading models already present in a simulator are likely sufficient for a first-order solution to a given problem. That is, even considering only the free-space path loss is better than completely ignoring any path loss in a simulation. If fading model availability does not guide the designer to choose a specific model, the constraints of the problem itself such as frequency, antenna heights, and propagation environment should bound the number of applicable fading models. But what if the needed fading model is not present in the designer's simulator of choice? First, let's continue with the mobile WiMAX theme set forth in the previous section with the ITU M.1225 channel model. Table 2-7 summarizes the mobile WiMAX channel models available as of this writing for NS-2, OPNET, and QualNET.

If the simulation designer's tool of choice does not offer the desired fading model, there are a few options available. First, the simulation designer may develop his or her own interpretation of a known fading model, such as the ITU M.1225. This requires a significant skill level in the native programming language of the simulator and is beyond the scope of this book. Another approach is to employ hardware-in-the-loop test methods with channel emulator hardware. This could be a customized solution built from commercial-off-the-shelf (COTS) components such as described in [69] or a commercial fading channel emulator such as those described in [100] and [101]. This is the approach taken by the WiMAX Forum for controlled conformance testing and generally offers the flexibility to test actual hardware with a configurable fading channel. The disadvantage to this approach is that channel emulators can be expensive, especially if multiple input, multiple output (MIMO) support is required and actual hardware or emulated hardware is required at the input and output of the channel emulation. A third approach is to use Matlab for processing the fading channel component of the model [102, 103], specifically leveraging the Communications Toolbox. The purpose of this section is not to provide the reader with an exhaustive description of the capabilities of Matlab's Communications Toolbox, but to describe how Matlab can facilitate the simulation of fading channel parameters.

```
% Six-Path Rayleigh channel with 60 km/hr velocity
v1 = 16.66;  %60 km/hr / 3600 sec (ITU Vehicular A velocity)
fc = 2E9;  %2 GHz carrier frequency

lambda = 3E8/fc; %c/f = lambda

fd1 = v1/lambda;

h = rayleighchan(1/(10*fd1), fd1, [0 310E-9 710E-9 1090E-9 1730E-9...
    210E-9], [0, -1, -9 -10 -15 -20]);
tx = randint(500, 1, 2);          % Random bit stream
dpskSig = dpskmod(tx, 2);         % DBPSK signal
h.StoreHistory = true;            % Allow states to be stored
y = filter(h, dpskSig);           % Run signal through channel
plot(h);
```

FIGURE 2-8. Modified Matlab example program that generates the ITU M.1225 Vehicular-A (60 km/hr) channel model.

Matlab and its graphical counterpart Simulink have emerged as two very important tools for digital signal processing, algorithm development, and even Monte Carlo simulation. Matlab's vector notation makes it natural for implementing channel models as tapped delay lines. The Communications Toolbox offers many tools to assist the designer in the development of fading channel models. One such tool is the Channel Visualization Tool, which allows the user to define a fading channel of type Rayleigh or Rician and pass a user-defined signal through the channel using the filter command. A modified version of the Matlab-provided Channel Visualization Tool example program is provided in Figure 2-8 that generates the ITU M.1225 Vehicular-A (60 km/hr) Channel Model. A Rayleigh fading channel is generated with a vector of tap delays and corresponding relative amplitude values taken directly from Table 2-6. The Doppler shift is calculated using Equation 2-7 directly and specified in the creation of the Rayleigh channel.

In this example, a random stream of 500 bits is modulated using differential binary phase shift keying (DBPSK) and sent through the channel using the 'filter()' command. The Channel Visualization Tool then allows the user to view the various characteristics of the channel in a GUI. Figure 2-9 illustrates one such characteristic of the channel, the Doppler spread, which matches the Doppler spread for the corresponding channel in Figure 2-7 that was generated using Equation 2-15. The output of the fading channel, 'y' in this case, could then be processed to generate a compatible input for a chosen simulator. For example, hard bit decisions could be made such that BER or FER metrics could be generated and accounted for before entering the higher layer simulation. Depending on the simulator, there may be other conditioning that could allow the baseband, time-domain data in the vector y to be fed into the simulator.

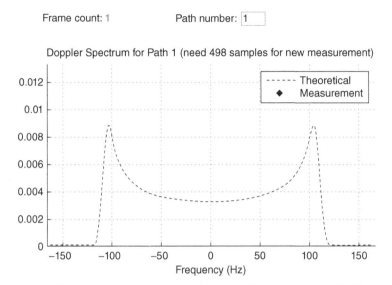

FIGURE 2-9. Classic Doppler spectra using Matlab Communications Toolbox for the ITU M.1225 Vehicular-A (60 km/hr) channel model.

This section purposely focused on the ITU M.1225 channel model due to its widespread acceptance and usage in Mobile WiMAX conformance testing. As noted in [94], there are likely some oversimplifications in this model that may have an impact on results for particular scenarios; however, simplifications, while not necessarily indicative of reality, are sometimes necessary to make the model tractable. Nevertheless, the reader should be aware of the limitations of these models in operating frequency and transmitter-receiver separation distance. Additionally, the interested reader that would like more information about other well-known cellular models not directly discussed in this book is referred to [94], where a concise summary of COST 207 and 231 model parameters, limitations, and usage cases is provided.

2.4 RF PROPAGATION SIMULATORS

As previously discussed, the four commonly used wireless network simulators offer various large-scale and small-scale propagation models to incorporate fading into simulations; however, some projects require much higher fidelity modeling of an electromagnetic environment. These software packages are most heavily used by wireless service providers and hence the most popular tool is that which is used by the most service providers. For example, a common application for a RF propagation software tool would be to determine the coverage area of a cellular network. It is no coincidence that significant propagation studies such as Hata and COST-Hata [20] have been conducted in the cellular bands such as 824 to 894 MHz and 1800 to 2000 MHz Personal Communications Systems (PCS) bands in North America, given the potential revenue generation from cellular customers.

Table 2-8 presents a survey of many of the major propagation tools and associated capabilities as a resource to the reader that seeks to investigate the various tools currently available [1].

The reader should keep in mind that, although some of these tools offer novel approaches to modeling the electromagnetic environment between a transmitter and receiver, many of these packages differ only in their user interface and visualization tools and not necessarily their choice of underlying propagation models. In general, there are two types of RF propagation models: empirical-based and physics-based. Most commercial products incorporate both types of models and allow the user to select which model to employ. Empirical models calculate large-scale path loss for a specified environment based on models that have been derived from curve fitting to empirically collected data. Examples of these empirical models include the well-known Okumura, Hata, and COST 207 and 231 models [7]. Physics-based models use ray tracing techniques to determine reflection, diffraction, and scattering estimates based on a three dimensional (3-D) environment. Standard datasets that characterize an area, such as Digital Terrain Elevation Data (DTED), are imported into these tools or the user creates a 3-D environment from scratch. The characteristics of all objects are considered including building materials to estimate path loss [7]. Three-dimensional ray tracing methods are becoming popular for modeling electromagnetic environments since they are not constrained by the limitations of empirical models. These limitations include the antenna heights and frequency ranges, which are a natural result of the heuristics based on statistical analysis of a given environment [21].

The authors do not necessarily have insight into the most popular RF propagation software being used by particular wireless service providers. However, as a case study to provide insight into general capabilities of these types of packages, Remcom's Wireless InSite package (http://www.remcom.com/) is described in detail in the following section. Instead of appealing strictly to the cellular telephony industry, Wireless InSite is also used in the research and development and military communities. This section provides details about Remcom's Wireless InSite package based on marketing material provided from the Remcom website as an overview of its capabilities.

2.4.1   Remcom's Wireless InSite

Remcom's Wireless InSite [22] is a software tool for modeling the effects of buildings (tunnel approximation is possible), terrain, and foliage on the propagation of electromagnetic waves. It is equipped with a graphical user interface (GUI), which allows the user to define the environment as accurately as the

TABLE 2-8. A Survey of RF Propagation M&S Tools

Tool Name	Company	URL
Wireless InSite*	Remcom, Inc,	http://www.remcom.com/wireless-insite/overview/ wireless-insite-overview.html
Atoll*	Forsk	http://www.forsk.com
Athena*	Wave Concepts	http://www.waveconceptsintl.com/athena.htm
CellOpt*	Actix	http://www.actix.com/main.html
Comstudy*	RadioSoft	http://www.radiosoft.com
EDX SignalPro*	EDX Wireless	http://www.edx.com/products/signalpro.html
ENTERPRISE Suite*	AIRCOM International	http://www.aircominternational.com/Software.html
LANPlanner*	Motorola, Inc.	http://www.motorola.com
Mentum Planet*	Mentum S.A.	http://www.mentum.com
NP WorkPlace*	Multiple Access Communications Ltd	http://www.macltd.com/np.php
Pathloss*	Contract Telecommunication Engineering	http://pathloss.com/
PlotPath*	V-Soft Communications LLC	http://www.v-soft.com/web/products.html
Probe*	V-Soft Communications LLC	http://www.v-soft.com/web/products.html
Profiler-eQ*	Equilateral Technologies	http://www.equilateral.com/products.html
RFCAD*	Sitesafe	http://www.rfcad.com/
RPS*	Radioplan GmbH	http://www.radioplan.com/products/rps/index.html
SEAMCAT* (free download)	European Communications Office	http://www.seamcat.org/
Volcano*	SIRADEL	http://www.siradel.com
Wavesight*	Wavecall	http://www.wavecall.com
WinProp*	AWE Communications	http://www.awe-communications.com/
Interactive Scenario Builder	US Naval Research Laboratory	https://builder.nrl.navy.mil/

* Indicates the platform is a commercial product.

user desires. This includes specifying transmitter and receiver locations, defining antennas, and signal characteristics such as transmit power and frequency. Wireless InSite contains many built-in antenna types, but data for a specific antenna may also be imported. The user can either utilize Wireless InSite's built-in database of building materials or define new materials through their electrical properties (electrical conductivity and relative permittivity).

Wireless InSite displays the environment (urban, terrain, or foliage) in a three-dimensional manner. Additionally, various widely used data formats such as DTED and AutoCAD can be easily imported into the Wireless InSite tool. The output data can be expressed in many different ways to fit the user's specific application, including received power, path loss, time of arrival for each propagation path, power delay profile, delay spread, electric field magnitude and phase, direction of arrival, BER, and carrier to interferer ratio [22]. Wireless InSite utilizes both empirical-based and ray tracing propagation models. The Wireless InSite empirical-based propagation models include the Hata and COST-Hata models. The 3-D ray tracing models include Urban Canyon, Fast-3D Urban, and Full-3D, all based on the Uniform Theory of Diffraction (UTD). There is also a Moving-Window Finite Difference Time Domain (FDTD) and Urban Canyon module [21, 22]. As an add-on product, Remcom offers a real-time module for Wireless InSite that uses methods to reduce the computational complexity of Wireless InSite's 3-D model without sacrificing the fidelity of the end result [21]. The interested reader is referred to [21, 22] for more information on Wireless InSite.

2.5   PROPAGATION AND FADING SIMULATIONS— LESSONS LEARNED

The intention of this chapter is to introduce the reader to the underlying mechanics of commonly used fading models and to link those models back to the GloMoSim, NS-2, OPNET, and QualNET simulators. As previously stated, more is not always better. This is a common theme throughout this book and more generally throughout the field of M&S. The designer must understand the problem at hand sufficiently to include just enough detail to make the results meaningful. Admittedly this is easier said than done, and there are certainly areas in M&S where one must draw from experience. The reader most likely has questions after reading this chapter, such as "Is the Friis free space path loss sufficient for an IEEE 802.11 WLAN simulation with 20 nodes?" Or, "Do I need to consider Rayleigh or Ricean fading in this case?" The answer is always that it depends on the designer's requirements and objectives. Are all of the 20 nodes within line of sight to one another? That is the first decision that determines whether the Friis equation and two-ray model as well as Ricean fading apply. If the answer is that a LOS scenario does apply, then the designer is recommended to use the Ricean small-scale fading and perhaps model both the Friis equation and the two-ray model and compare the two results. NLOS scenarios should make use of the lognormal shadowing large-scale fading model and Rayleigh small-scale model. Perhaps the scenario to be modeled is a point-to-point microwave link with complete LOS in a rural scenario. In this case, accurate results can most likely be obtained while ignoring the small-scale fading effects. As discussed in Section 2.1.1, the choice of large-scale model depends on the designer's link-budget margin within the accuracy of the model. If a margin of 50 dB can be tolerated in the path loss, then the Friis equation is probably sufficient. Keep in mind that the error margins are generally quite large in propagation modeling due to the number of unknown environmental variables. From the author's experience, models that agree within 10 dB of empirical data are considered to be quite accurate. Therefore, the designer should not expect simulated results to exactly match empirical data collected in the field.

It is assumed that the majority of readers will apply the material from this chapter to one of the four commonly used network simulator models or one of the other network simulators that contains similar support for fading models. In this case, the designer has the ability to switch between fading models easily and is encouraged to do so. The reader is also encouraged to use Matlab and Simulink as self-educating tools to better understand the impact of fading models that are being incorporated into a simulation for the first time. This will help the simulation designer to better troubleshoot and validate his or her simulation; the field of M&S continues to mature, but models are still not perfect. The designer should have sufficient understanding of the particular scenario being modeled such that he or she can validate the result. That is, a strange result is most likely not a novel finding but an error in the simulator, as seen in multiple papers similar to [2, 4] that compare the accuracy of simulation results. This is an important lesson that especially applies to RF propagation simulators. The tools listed in Table 2-8 offer many different features, but few have exhaustively tested the possible simulation scenarios that may be created. In some cases, the newer versions of software may add new capabilities, but are also less stable than previous versions. Some capabilities of the older software releases may even be lost in newer releases.

Many of the RF propagation simulators are commercial products and it is recommended that modelers make use of the technical support provided with these products. Furthermore, the designer should expect to require technical support along the way, at least while becoming familiar with a simulator's capabilities. Three of the most common pitfalls for RF propagation software include the following: excessive computational time, unspecified boundaries, and general lack of documentation. Computational time for propagation simulators is often difficult to predict. Experience has shown that for some scenarios, a simulation of multiple receive nodes will run faster than for a single receive node. It is generally recommended as a rule of thumb that a designer's estimated calculation on the runtime of a given propagation simulation be doubled to account for any extra required computation time. Many simulators suffer from scant or vaguely written literature. This is one reason why technical support can be so helpful. The limitations and boundaries of underlying propagation models inside the simulator are not always provided to the user. For example, some models have boundary conditions placed on the frequencies or separation distances between nodes. In some cases the simulators may allow the user to define scenarios that surpass these bounds without warnings. Of course this makes any simulator output under these conditions questionable at best and likely invalid. In some cases the user may receive an error message, but this error may occur at the end of the simulation. For example, some ray tracing software will allow the users to specify a scenario that exceeds the maximum number of reflective surfaces allowed, yet the software only produces an error message once it has exceeded its memory allocation. The user should not be surprised if the simulator crashes; general programming guidelines apply in that the simulation will rarely produce the desired output from the first implementation. There is also a general lack of guidance in most simulators to help users choose a correct model appropriate to a particular scenario. The user must rely on previous experience, technical support, and third-party references to apply the correct models to particular scenarios.

As a summary of lessons learned:

- The simulation designer should understand and validate the model output.
- The application of particular propagation models to a given problem is scenario dependent.
- The simulation designer must not exceed the boundary conditions of a particular model applied to a scenario.
- Worst case computational time should be anticipated by doubling the designer's estimated run time.
- The designer should expect to make use of all technical support and third-party references available since simulation literature is limited.
- Run RF propagation simulators on a platform with the largest processor and random access memory (RAM) available.

CHAPTER 3

Physical Layer Modeling and Simulation

This chapter describes particular areas related to modeling the PHY in wireless networks that perhaps would not otherwise receive consideration from the designer. The topic of wired PHYs such as cable modems or Asynchronous Digital Subscriber Lines (ADSL) are omitted from this chapter. Although there are open research topics concerning interference mitigation, feedback, and equalization of these links, these are not strictly simulation problems and simulation results for wired systems are not frequently encountered in open literature. The topic of modeling the PHY in wireless networks is presented, motivated by the need to improve simulation practices and introduce necessary realism into models to improve confidence in simulation results.

This chapter presents some of the fundamental wireless PHY concepts to be considered in simulation design. The PHY components supported by the common simulators NS-2, OPNET, GloMoSim, and QualNET are also presented. Four NS-2 wireless PHY simulation examples are included to further reinforce the concepts described. This chapter concludes with suggestions to the reader by reviewing some of the lessons learned and common pitfalls of wireless PHY network simulation and corresponding mitigation strategies.

While many chapters in this book are written to address a particular layer of the protocol stack, the chapters are correlated with respect to a full wireless network simulation; Chapter 4 presents related wireless MAC topics that correspond to the PHY material presented in this chapter. Additionally, the RF propagation topics presented in Chapter 2 are highly correlated to the material presented in this chapter and in most cases could be considered as an extension of the general PHY topic. As shown in Figure 3-1, the propagation topics discussed in Chapter 2 can be considered as inputs to the PHY M&S block in Figure 3-1 and the wireless MAC topics are related to the output of the PHY M&S block. The PHY M&S block can be considered as

An Introduction to Network Modeling and Simulation for the Practicing Engineer, First Edition. Jack Burbank, William Kasch, Jon Ward.

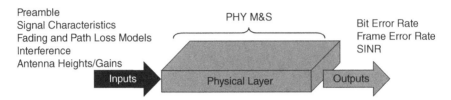

FIGURE 3-1. Wireless network PHY simulation example.

the underlying digital bits and corresponding analog waveforms that traverse a wireless link.

Indeed there are many considerations at the PHY including length of preamble, path loss, SINR, and fading delay; however, the impact of PHY parameters on simulation results is probably the least considered in published network simulation results. In fact, this is not problematic as long as the underlying PHY assumptions and corresponding parameters are well understood by the implementer and documented sufficiently with any published results. Especially in the case where upper-layer network protocols are being compared, as long as all protocols under comparison experience the same PHY impacts, then results are likely valid. The area of concern for the implementer is when the PHY and associated assumptions directly affect simulation results and when distinct scenarios are compared. In this case, the designer must verify that the assumed PHY parameters match the real-world scenario to which they are applied. For example, if a given simulation is to compare ad hoc routing protocols for a deep fading scenario in low SINR environments, the underlying PHY parameters must represent this scenario and not a scenario with high SINR.

The general PHY topics in this chapter are organized as follows: Incorporating Interference into a Model (Section 3.1), Importance of a Preamble (Section 3.2), Practical Wireless PHY Model Implementations (Section 3.3), and Wireless Network Simulation Common Pitfalls and Mitigation Strategies—PHY Layer (Section 3.4).

3.1 INCORPORATING INTERFERENCE INTO A MODEL

Interference exists in all wireless networks and its effects must be carefully considered in simulations. In many technologies, each mobile user must compete for the wireless medium, which is why the particular MAC being employed (e.g., ALOHA, Carrier Sense Multiple Access/Collision Avoidance (CSMA/CA), etc.) is so important in accurately simulating network performance; however, the interference from competing mobile users is not the only source of interference to be considered. The PHY is impacted by interference and noise that exist in the normal RF environment. A college professor once said that noise is the reason that communications engineers are employed. While noise is not the only component that degrades the quality of communications, the noise environment in which a given wireless system operates must be carefully considered. Besides the natural thermal noise that occurs over a system's RF operating bandwidth, there are also unintentional and intentional emissions with which wireless networks must directly compete. Consider the 2.4-GHz unlicensed ISM band where IEEE 802.11b/g networks operate. A wireless IEEE 802.11 mobile user operating at 2.4 GHz must compete with other IEEE 802.11 nodes—that may have dissimilar PHYs— and it must operate in the presence of microwave oven emissions, baby monitors, Bluetooth, IEEE 802.15.4 radios, and any malicious emissions such as jammers. The purpose of the IEEE 802.11 MAC is to coordinate between other competing IEEE 802.11 users, but other aforementioned sources of interference directly impact the PHY and therefore the performance of the higher network layers; the effects of interference and noise on the performance of the PHY must not be ignored in a wireless network simulation.

There is no universal method for incorporating interference into a wireless network simulation. In fact, an entire book could be focused on this single topic, as is much of [31], which includes discussions of noise equivalent bandwidth, quasi-analytical methods for determining BER, and other interference and noise-related topics. The reader is encouraged to investigate [31] and [32] for a more comprehensive treatment of methods for incorporating interference into simulations and determining BER. In the more practical sense, Figure 3-2 contains an example block diagram of an IEEE 802.11 superheterodyne transceiver architecture that transmits the 1 and 2 Mbps IEEE 802.11 DSSS waveform that is spread by a Barker sequence.

Certainly a model used to evaluate the performance of this transceiver would implement the intermediate frequency (IF) and RF stages and consider

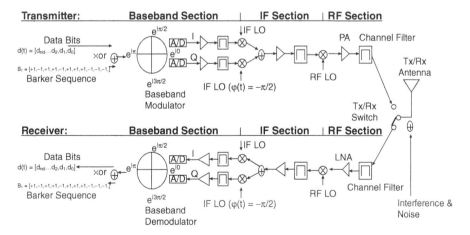

FIGURE 3-2. Full RF IEEE 802.11 simulation block diagram (1 and 2 Mbps).

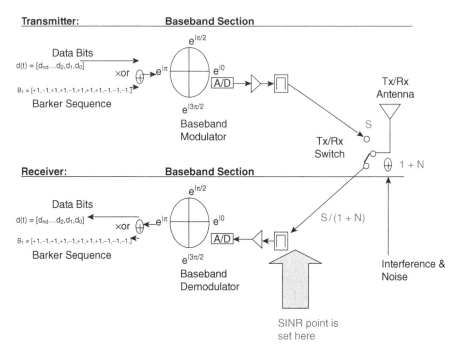

FIGURE 3-3. Baseband IEEE 802.11 simulation block diagram (1 and 2 Mbps).

any conversion losses, intermodulation products, or oversampling and under-sampling degradations; however, this is not the goal of most wireless network simulations and not the goal of this section. In fact, the IF and RF stages of the block diagram in Figure 3-2 are largely irrelevant to the performance of higher layer networking protocols since most academic papers assume system linearity and perfect up and down conversion. Therefore, the simplified block diagram shown in Figure 3-3 is a better representation of the PHY of a wireless network simulation in a common simulator such as NS-2 or a Simulink model.

In Figure 3-3, only the baseband section of the system is implemented. That is, the data bits are spread with the Barker sequence and modulated at baseband into complex numbers according to the modulation type chosen (e.g., binary phase shift keying (BPSK) or quadrature phase shift keying (QPSK)). Filtering and amplification may be done next, but this is usually ignored in this type of simulation since the performance is being set by the noise and interference level at the antenna input and this noise component is assumed to be band-limited to the operation band of the communication system being modeled. The noise-free output of the transmitter is the desired signal (S), and S would be the received signal in a simulation scenario that considers an ideal channel. The interference (I) and noise (N) components are added directly to

S to set the SINR (S/(I + N)). This is designated in Figure 3-3, where the SINR point is calculated as given in Equation 3-1.

$$SINR = \frac{P_R}{N_0 + \sum_k P_k} \qquad (3\text{-}1)$$

In Equation 3-1, P_R represents the received power level of the desired user, which already accounts for any path loss or signal distortion effects between the desired transmitter and receiver. P_k represents the received signal power of the kth interferer and N_0 represents the two-sided power spectral density of the environmental thermal noise. Each sample that passes through the analog to digital converter (A/D) contains the contributions of S + I + N and are fed into a baseband demodulator and the data bits are then recovered through despreading in the case of the example DSSS IEEE 802.11 system. The model implementer chooses the interference and noise levels to set a target SINR to simulate the desired scenario. Depending on the noise and interference environment of interest, an average SINR target can be set in the simulation. In addition to the path loss and antenna losses, which are generally accounted for in Equation 3-1, additional sources of loss such as the noise figure of any front-end receiver components (e.g., low noise amplifiers (LNAs)) should be considered. Consider the propagation environment characterized by the free space path loss FS given by Equation 2-1. In this case, a link budget equation representing the kth user's received power is given by Equation 3-2, where NF(dB) accounts for the desired receiver's noise figure and other losses [106].

$$P_k(dB) = P_t(k) + G_t(k) + G_r - FS(dB) - NF(dB) \qquad (3\text{-}2)$$

The noise component in a wireless simulation is almost always chosen to be additive white Gaussian noise (AWGN). This serves two purposes in academic research. First, the Gaussian distribution provides a nice closed-form, tractable, analytical solution that allows simulation results to be validated through analytical solutions. Second, and more importantly, the sum of a sufficiently large number of independent and identically distributed (i.i.d) random variables approaches a Gaussian distribution by invoking the central limit theorem. By definition AWGN is independent of the desired signal and, in a very heavy interference environment, interferers can often be modeled as Gaussian distributed. In this case, the interference and noise components I + N collapse to a single AWGN component N, with mean and variance chosen to set the SINR. A Gaussian distributed random variable is completely defined by its mean and variance, where a larger noise variance sets a smaller SINR and a smaller noise variance sets a higher SINR. This means that noise and a sufficiently large amount of multi-user interference may be chosen in a simulation as samples of Gaussian distributed random variables with a given mean and variance, most commonly chosen as zero mean and variance of N_0.

3.1.1 Iterative Summation of Interferers

As discussed in Chapter 2, fading models may be applied iteratively to individual interferers to determine the interference incident on a given desired receiver node as an alternative to making simplifying assumptions that interference is Gaussian distributed. This follows directly from the example presented in Section 2.3, where the contributions of an interferer, including the large-scale and small-scale fading effects incurred while traveling to the receiver, are simulated by passing a signal through an impulse or frequency response model. In this case, the computational resources required at the simulated receiver can be significant since the interference summation must be computed iteratively for each transmitted data bit and applied to the desired signal before detection [35]. Additionally, the phase shifts and delays associated with the multi-user interference components must be accounted for, since they will both constructively and destructively affect the SINR calculation. Generally, even in the case of a simulation that sums over multiple interferers, thermal noise is incorporated into the system by way of a Gaussian random variable with sample mean of zero and variance of N_0. In this case, the N_0 that represents thermal noise is significantly less than the N_0 that would represent multi-user interference.

The limited interference model is often applied to iterative multi-user interference simulations to reduce computational complexity by setting a threshold limit on the propagation range of interfering signals and therefore limiting the amount of program iterations required. Essentially, this method assumes that beyond a given threshold, the contributions of interferers are significantly small compared to other louder interferers and hence can be discarded without significantly degrading the calculation [35]. By discarding interferers, the number of summations reduces and the model should scale better.

The difficult problem for the simulation designer wishing to employ limited interference is to determine a satisfactory threshold. Unfortunately, this is largely a trial-and-error problem that is completely dependent on the scenario being modeled. The designer should begin with determining the greatest path loss experienced between the furthest two nodes in the network using one of the large-scale models discussed in Chapter 2 and begin setting the threshold experimentally below this data point. Depending on the specific metrics being modeled, trial runs of the simulation can spot check the point at which the threshold drops too low and errors are introduced into the output. The interested reader is referred to [35], which contains an example experiment to which the authors applied the limited interference model to determine the impact of two different thresholds.

TABLE 3-1. PHY models for SNR available in GloMoSim, NS-2, and OPNET. Recreated from [10]

Simulator Parameter	GloMoSim (v.2.02)	NS-2 (v. 2.1b8)	OPNET
Noise (SNR) calculation	Cumulative	Comparison of two signals	Cumulative
Signal reception	SNRT based, BER based	SNRT based	BER based

3.1.2   Calculation of SINR and FER

Once the noise and interference contributions are incorporated into a simulation, by way of a Gaussian distributed random variable or iterative summation of interferer contributions, the SINR must be chosen at the desired receiver. This is the metric the simulated receiver shall use to determine whether or not a received signal should be further processed. There are generally two common methods for calculating SNR in a simulation: 1) SNR threshold (SNRT)-based and BER-based. Table 3-1 lists the methods used by common simulators such as NS-2, GloMoSim, and OPNET that calculate interference and noise at the desired receiver [10].

The SNR threshold (SNRT) method simply compares the SNR of each received sample at the SINR point shown in Figure 3-3 and compares it to a threshold. That is, if the instantaneous SNR sample is greater than or equal to the SNRT, the sample is sent into the demodulator and decoder. Otherwise, if the sample is less than the threshold, it is assumed too noisy to be decoded and is discarded.

As with any engineering problem with multiple solutions, there exist tradeoffs between these two solutions. First, the SNRT solution is not very realistic. In an actual receiver implementation, unless the received signal $(S + I + N)$ is below the receiver sensitivity, it would generally be passed to the demodulator and decoder and show up in error during upper-layer processing; however, in a simulation, this simplistic method provides computational savings compared with the BER-based solution [10]. Additionally, the SNRT method does not consider forward error correction (FEC) as is applied to many communication systems. In many cases the FEC can recover data bits at the decoder even in low SNR environments. In a practical example, the SNRT method for calculating interference and noise contributions would handle an IEEE 802.11b and IEEE 802.11g system equivalently without accounting for the FEC applied to the IEEE 802.11g signal.

The BER-based solution extrapolates frame error rate (FER) from BER, which is determined probabilistically at a given SNR for a specific modulation and FEC type. The BER-based method is more realistic than the SNRT since the FER accounts for a cumulative effect of received signal variation and because FEC is considered in this method; however, a derivation of FER from

BER may not always be accurate for a wireless channel with significant burst errors, where bit errors are no longer independent. The general equation for FER (Equation 3-3) assumes bit errors are independent [107] where n denotes the number of bits in an average frame.

$$FER = 1 - (1 - BER)^n \tag{3-3}$$

Figure 3-4 shows a plot of BER for various uncoded PSK modulation types versus Eb/N0 [33]. Eb/N0 is the energy per bit divided by the AWGN noise density and is easily converted to SNR by Equation 3-4.

$$SNR = \frac{E_b}{N_0} + 10 \log_{10}(\text{DataRate}) \tag{3-4}$$

Note that the BER plots shown in Figure 3-4 are well known and available for various common modulation types in the cases of both uncoded and coded modulation. In the coded case, a variety of FECs are applied to the system and the graphs shift to the left, depending on the exact code applied. That is, a coded system improves the BER and corresponding FER achieved at a lower Eb/N0.

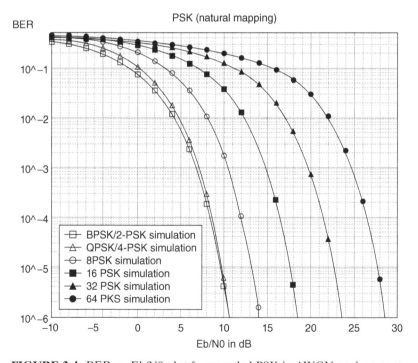

FIGURE 3-4. BER vs. Eb/N0 plot for uncoded PSK in AWGN environment.

A coded system may also improve a wireless receiver's ability to recover from burst errors in the channel. Especially when modeling fading channels as discussed in Chapter 2, a consequence of modeling time-varying channels and fluctuations in signal amplitude at the receiver is an increase in bit errors. Wired networks may achieve textbook BERs of 10^{-6} or better compared with a wireless environment where a BER of 10^{-3} may be considered very good, depending on the multipath conditions, multi-user interference, and ambient interference. A BER or FER of 10^{-3} for most simulations is relatively easy to simulate quickly, but as target BER or FER decreases, the computational time required for simulation increases. As a rule of thumb, the simulation designer should run a minimum number of trials N to equal an order of magnitude larger than the inverse of the BER (e.g., $N = 10 \times (1/BER)$) [32]. An example for the case of a BER of 10^{-3} is given in Equation 3-5 [32].

$$\text{Desired BER} = 10^{-3}, \text{N} = 10 \times \left(\frac{1}{10^{-3}} \right) = 10^4 \text{ trials} \qquad (3\text{-}5)$$

The FER can then be computed from the simulated BER results by way of Equation 3-4, under the assumption that data bits are independent. These so-called Monte Carlo BER simulations are among the cases in simulation where more is better. Recall that a BER or FER of 10^{-3} is one error in one thousand, so we should expect approximately ten errors in ten thousand; however, it is always best to run the most number of trials possible based on the execution time and due date for results.

3.2   THE IMPORTANCE OF A PREAMBLE

In a digital communications system, a preamble is a known sequence of bits that are transmitted to delimit frame starting positions and can be used for time and frequency synchronization. Contrary to the section title, there are instances in simulation where the preamble may be ignored with no conse-quence to the output. For example, a Monte Carlo simulation that seeks to investigate only the BER of a BPSK threshold detector for various levels of SNR in an AWGN scenario need not consider the preamble. In this case, a simplified system model should consider all bits as equally likely, with the order of the bits denoting no additional information. There is no synchroniza-tion circuit in this PHY simulation since the detector is clocked based on the arrival of each raw bit and a detection decision is determined bit-by-bit. Additionally, this simple model contains no MAC above the PHY to make channel access decisions, and hence no preamble is needed to identify one signal type from another. Compared with the simple Monte Carlo BPSK detector, the preamble is especially important in low SINR conditions, where the known preamble can be used by the receiver to determine the existence of an intelligible signal. Consider the case where an IEEE 802.11 MAC is being modeled that implements virtual carrier sensing when reserving the channel. In this case, the preamble must be present to identify an IEEE 802.11 signal from other interfering signals; in this case, the preamble bits cannot be considered the same as arbitrary payload bits.

The preamble is important for frequency-hopping radios and future systems that rely on dynamic spectral access (DSA) at the PHY to occupy unused spectrum. In this case, the receiver must synchronize to the transmitter and each newly chosen frequency uses the known preamble sequence. The preamble may also be used in MIMO systems to allow synchronization such that the diversity gains from multiple spatial receivers can be coherently combined. Many simulation designers take the approach of packetizing data bits with appropriate header information at the higher layers of the protocol stack and considering everything below the layer of interest as simply data bits or payload data. While in some cases, such as the previously described BPSK BER example, this abstraction may be sufficient, this is not generally a valid assumption. If the goal of the simulation is to evaluate a parameter related to the PHY or MAC layer, the preamble should not be considered as simply additional payload bits.

In short, preamble bits are not just data bits, they serve additional purposes in signal identification and synchronization. In IEEE 802.16-2004 [37] networks, the preamble is used for synchronization purposes in denoting the transition between the uplink (UL) and DL portions of frames in the general Time Division Duplex (TDD) system. The DL and UL frame structure containing preambles for the IEEE 802.16-2004 system are illustrated in Figure 3-5. Because the IEEE 802.16-2004 system PHY is based on an orthogonal frequency-division multiplexing (OFDM) waveform, the preamble is specified in terms of OFDM symbols and not bits. In the general case of a data payload, the data bits are mapped to OFDM symbol subcarriers via the inverse fast Fourier transform (IFFT); however, in the case of the preamble, the receiver demodulator only cares about the underlying OFDM symbol received at the correct time and hence no data bits modulate the preamble. IEEE 802.16-2004 BTSs may be synchronized to a global time standard such as Global Position Satellite system (GPS), but in the more general case, all subscriber nodes must synchronize their time and frequency reference with the BTS via the DL preamble. Hence, a simulation of the IEEE 802.16-2004 system that does not consider the preambles does not properly implement the system's synchronization methods and may not provide accurate results, even when evaluating higher layer metrics.

IEEE 802.11b systems as specified in [17] use a 144-bit preamble that contains a 128-bit Sync field and a 16-bit Stop Frame Delimiter (SFD) field shown in Figure 3-6. The Sync field contains all ones and the SFD field contains the binary string 1111 0011 1010 0000 transmitted rightmost bit first. The IEEE 802.11b DSSS system preamble and entire protocol data unit (PDU) is chipped

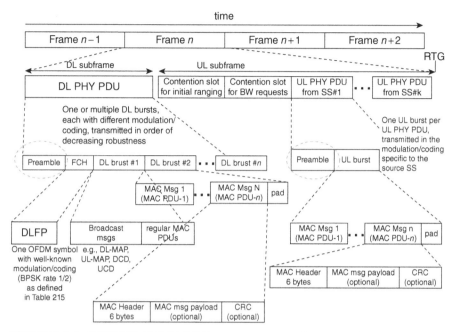

FIGURE 3-5. IEEE 802.16-2004 TDD frame structure demonstrating the preamble [37].

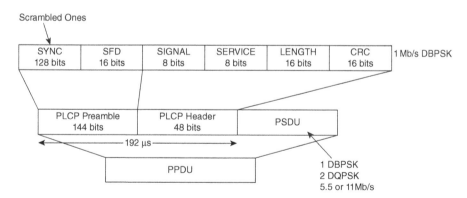

FIGURE 3-6. IEEE 802.11b HR DSSS PLCP long frame format [17].

with a high-rate Barker sequence, so there is a concept of the preamble being composed of actual bits. This is not the case with IEEE 802.11a and g systems that are based on OFDM waveforms and use a preamble structure much like that of IEEE 802.16-2004. The long physical layer convergence protocol (PLCP) preamble and header are always transmitted using DBPSK modulation at the 1 Mbps data rate.

The simulation designer that seeks to measure system application through-put must also consider the amount of overhead consumed by PHY preambles. Depending on the signal structure and the data rate, the preamble may compose a significant amount of a frame's bits that are no longer available for higher layer payload data. As an example, the amount of overhead added to an IEEE 802.11b frame by the preamble (long header) is 192 bits, which can be significant overhead at lower data rates. As the frame size increases, the percentage of transmit time occupied by the preamble overhead significantly decreases, but for small frame sizes the preamble transmission overwhelmingly dominates the channel occupation time. In a generic wireless network simulation, perhaps a shorter preamble could be used with minimal impact to the system performance metrics under evaluation; however, an accurate IEEE 802.11b simulation that attempts to capture the true performance, in terms of higher-layer or MAC-layer throughput or delay, must consider the preamble overhead in terms of both the extra channel occupation incurred and the impact of preamble synchronization on the system. The impact of the IEEE 802.11b preamble length on application throughput is considered in an example in Section 3.3 using NS-2.

3.3 PRACTICAL WIRELESS PHY MODEL IMPLEMENTATIONS

It is the authors' intention to provide the reader with a set of guidelines to improve the quality of his or her simulations. Since the audience is composed of practicing engineers, example reference models should accompany topics presented; however, the possible number of even basic examples from the popular network simulators described in this book could easily fill multiple books. In an attempt to balance utility and brevity of examples, the authors chose to only focus on the NS-2 simulator. This is motivated by a few reasons: 1) NS-2 is one of the most widely used simulators in academia; 2) NS-2 is freely available for any reader to download, install, and recreate presented examples; 3) the NS-2 implementation of WiMAX has received WiMAX forum's endorsement [99]; and 4) the open-source format allows the reader to further investigate topics in as much detail as he or she chooses.

The authors' usage of NS-2 should not be interpreted as an endorsement for this simulator. In fact, the authors' have no allegiance to a single M&S platform. Along those lines, a few words about NS-2 are in order. NS-2 provides a great learning environment because of the open access to both the C++ libraries and Tcl scripts, but the learning curve is quite steep. From the installation process to running a pre-made simulation example can take many hours. The installation process can be quite temperamental, depending on the distribution of Linux used for installation. The authors used NS-2 version 2.31 on a Fedora release 8 (Werewolf) Linux system for the NS-2 simulation examples in Chapters 3 and 4 [110]. The purpose of this section is to describe best practices as applied to a NS-2 simulation example and not to describe how to use NS-2. Additionally, examples consider only NS-2 wireless simulations.

Depending on the parameters set in the Tcl script, NS-2 may provide multiple output formats. In general, there are two standard NS-2 output types: 1) a .nam file that visualizes node locations when input to the Network Animator NS-2 add-on package and 2) an event trace file. Note that there are two NS-2 trace file types for wireless simulations, referred to as the "old" and "new" trace files [111]. The trace files are of file extension .tr and contain timestamped text entries for each simulation event. The authors are unaware of any standard tool for processing these files; researchers generally use their favorite scripting language such as Perl or AWK to process the trace files with customized scripts. Processed data may then be visualized through plotting in programs such as Matlab, GNU Plot, or Excel.

The authors chose NS-2 examples from both IEEE 802.11 and IEEE 802.16 technologies because these are two NS-2 models available and they are well aligned with topics discussed throughout this book. The first series of PHY examples considers varying PHY parameters in an IEEE 802.11b simulation and the corresponding impact on application-layer throughput. The second example illustrates the average application throughput achieved when simulating an IEEE 802.16-2004 system using the NS-2 National Institute of Standards & Technology (NIST) model [113] for various data rates and cyclic prefixes (CPs).

3.3.1 NS-2 IEEE 802.11b Wireless PHY Example

To illustrate the importance of correctly simulating the PHY layer for a given simulation scenario, three simple, two-node IEEE 802.11b examples are provided, based on the frequently referenced NS-2 "simple-wireless.tcl" example [99]. In each example, a PHY parameter is varied and the impact to system parameters such as throughput is observed by comparison with a baseline wireless configuration. The objective here is not to demonstrate whether or not the NS-2 IEEE 802.11 PHY implementation represents a real-world scenario, but rather to illustrate that the NS-2 output is completely dependent on the PHY parameters chosen by the simulation designer. When these parameters are not explicitly defined in the NS-2 Tcl file, default parameters are used that may not match the actual scenario under consideration. The authors purposely chose these examples from [109] such that the Tcl scripts are available for readers to download and investigate further. The three IEEE 802.11 wireless simulation parameters considered are summarized in Table 3-2.

The authors chose parameters in examples 1 and 2 directly from [17]; because the parameters in Example 3 depend on the geometry of the particular scenario being simulated, CSThreshold_ and RXThreshold_ were arbitrarily chosen to be possible real-world values for an indoor IEEE 802.11 scenario.

TABLE 3-2. Summary of NS-2 IEEE 802.11b PHY Parameters Varied in Each Simulation Example

Example 1: Data Rate	Example 2: Preamble Length	Example 3: SNR Thresholds
Baseline: 1 Mbps Comparison: 11 Mbps	Baseline: 144 bit (long) Comparison: 72 bit (short)	Baseline: CSThreshold_ = 5e-11, RXThreshold_ = 5e-8 Comparison: CSThreshold_ = 5e-12, RXThreshold_ = 5e-9

TABLE 3-3. IEEE 802.11b Data Rate PHY Example Parameters

Baseline	Comparison
`Mac/802_11 set dataRate_ 1Mb` `Phy/WirelessPhy set` `  bandwidth_ 1e6`	`Mac/802_11 set dataRate_ 11Mb` `Phy/WirelessPhy set` `  bandwidth_ 11e6`

3.3.1.1 IEEE 802.11b Data Rate PHY Example with NS-2

Most IEEE 802.11b networks use adaptive modulation such that control, data, and management traffic may be sent at varying data rates depending on channel conditions. NS-2's baseline package does not contain adaptive mechanisms, but rather the data rates of all wireless transmissions are defined by the user. The purpose of this example is to demonstrate what should be an intuitive result, that the chosen data rate has a considerable effect on the application-layer throughput of the system. Table 3-3 illustrates the changes made to the "simple-wireless.tcl" simulation description, where in each case the two commands are added to the top of the simulation before **set val()** statements are initialized [109].

Table 3-4 summarizes a subset of possible output parameters processed by applying a Perl script to the NS-2 trace file, where the 1 Mbps case is shown on top and the 11 Mbps case on the bottom. When using the 1 Mbps data rate, approximately one third of the amount of data packets are sent compared to the 11 Mbps case. This corresponds to an approximate one third decrease in application-layer throughput as well. In this example, the 144 bit long IEEE 802.11b preamble is used for both data rate simulations. If the simulated data rate were not set to correctly match the IEEE 802.11b equipment being used in the real-world scenario, any throughput predictions would likely be invalid.

3.3.1.2 IEEE 802.11b Preamble PHY Example with NS-2

Section 3.2 describes the importance of correctly defining the preamble in a simulation

TABLE 3-4. IEEE 802.11b Data Rate PHY Example Output

```
====== Analyzing NS-2 Wireless Data (1Mbps)======
Data Packets Sent          : 2122 pkts (1155880 B)
Data Packets Received      : 2091 pkts (1170480 B)
Simulation Duration        : 149.48989626 s
Avg. Application Throughput : 385293.333333333 (bps)

====== Analyzing NS-2 Wireless Data (11Mbps)======
Data Packets Sent          : 6915 pkts (3737600 B)
Data Packets Received      : 6883 pkts (3861000 B)
Simulation Duration        : 147.926068772
Avg. Application Throughput : 1196032 (bps)
```

TABLE 3-5. IEEE 802.11b Preamble PHY Example Parameters

Baseline	Comparison
Mac/802_11 set	Mac/802_11 set
PreambleLength_ 144	PreambleLength_ 72

TABLE 3-6. IEEE 802.11b Preamble PHY Example Output

```
====== Analyzing NS-2 Wireless Data (144bit preamble)======
Data Packets Sent          : 2122 pkts (1155880 B)
Data Packets Received      : 2091 pkts (1170480 B)
Simulation Duration        : 149.48989626 s
Avg. Application Throughput : 385293.333333333 (bps)

====== Analyzing NS-2 Wireless Data (72 bit preamble)======
Data Packets Sent          : 2225 pkts (1210000 B)
Data Packets Received      : 2195 pkts (1229720 B)
Simulation Duration        : 143.979344102 s
Avg. Application Throughput : 420869.565217391 (bps)
```

and not treating preamble bits as simply a longer string of data bits. This example demonstrates that the specified length of the IEEE 802.11b preamble, either 144 bits (long) or 72 bits (short) affects the application-layer throughput of the system. Table 3-5 illustrates the changes made to the "simple-wireless. tcl" simulation description in this example.

Table 3-6 summarizes a subset of possible output parameters processed by applying a Perl script to the NS-2 trace file, where the 144 bit (long) preamble case is shown on top and the 72 bit preamble case on the bottom. Both cases use a data rate of 1 Mbps. Over the simulation time of approximately 150 s, the

TABLE 3-7. IEEE 802.11b SNR Threshold PHY Example Parameters

Baseline	Comparison
Phy/WirelessPhy set CSThresh_ 5e-11	Phy/WirelessPhy set CSThresh_ 5e-12
Phy/WirelessPhy set RXThresh_ 5e-8	Phy/WirelessPhy set RXThresh_ 5e-9

data throughput corresponding to the 72 bit preamble case is approximately eight percent greater than the 144 bit preamble case (420.869 kbps compared with 385.293 kbps). Depending on the margin of error that is acceptable to the simulation designer, the effects of the preamble may or may not make throughput predictions invalid.

3.3.1.3 IEEE 802.11b SNR Threshold PHY Example with NS-2
Section 3.1 describes the different methods for incorporating interference into a simulation, focusing on generic principles that could be employed most easily by the developer in a homebrew simulation. As noted in Table 3-1, NS-2 uses the SNR threshold comparison method to handle interference. These thresholds are set using two parameters: the CSThreshold_ and the RXThreshold_. The CSThreshold is used to determine whether a signal is present at the desired receiver; that is, does the received power level exceed the CSThreshold? If the signal power does not exceed CSThreshold, it is considered a dropped frame. If the signal exceeds the CSThreshold, then it is compared with the RXThreshold. A signal that exceeds RXThreshold is considered a valid frame (e.g., no CRC error). Otherwise, the packet is marked in error. As previously noted, these threshold values are completely scenario dependent and should be set according to the expected value of real-world measurement data. Table 3-7 illustrates the changes made to the "simple-wireless.tcl" simulation description in this example.

Table 3-8 summarizes a subset of possible output parameters processed by applying a Perl script to the NS-2 trace file, where the higher threshold value case is shown on top and the lower threshold value case on the bottom. Both cases use a data rate of 1 Mbps and a 144 bit (long) IEEE 802.11b preamble. Over the simulation time of approximately 150 s, the data throughput corresponding to the lower threshold value case is approximately 22% greater than the higher threshold value case (493.182 kbps compared to 385.293 kbps). This difference in throughput only represents a received power threshold difference of an order of magnitude, or about 10 dB. As previously mentioned in Chapter 2, in the field of propagation prediction, variations of 10 dB are considered relatively minor. Although a 22% margin of error may be acceptable to the simulation designer, consider the cumulative effects of altering a threshold value by many orders of magnitude.

TABLE 3-8. IEEE 802.11b SNR Threshold PHY Example Output

```
====== Analyzing NS-2 Wireless Data (CST = 5e-11,RXT =
  5e-8)======
Data Packets Sent             : 2122 (1155880 B)
Data Packets Received         : 2091 (1170480 B)
Simulation Duration           : 149.48989626s
Avg. Application Throughput    : 385293.333333333 (bps)

====== Analyzing NS-2 Wireless Data (CST = 5e-12,RXT =
  5e-9)======
Data Packets Sent             : 4108pkts (2219320 B)
Data Packets Received         : 4078pkts (2291720 B)
Simulation Duration           : 149.014447724s
Avg. Application Throughput    : 493182.222222222 (bps)
```

TABLE 3-9. Summary of WiMAX PHY Features Implemented in Common Network Simulators. Recreated from [108]

Feature	QualNET [97]	OPNET [98]	NS-2 NIST* [113]
Standard	IEEE 802.16e	IEEE 802.16e	IEEE 802.16e
PHY	OFDMA	OFDMA	OFDM
AMC	No	No	No
BER	No	No	No

*Denotes this model and support as being provided by user community.

The three simulation examples demonstrate that varying a single IEEE 802.11 PHY parameter may cause a significant change in the simulation output. This should further reinforce the need to carefully choose input parameters to the model that are applicable to the real-world scenario of interest.

3.3.2   NS-2 WiMAX PHY Example (IEEE 802.16 NIST Model)

WiMAX has recently become of interest to M&S researchers as it becomes an increasingly deployed Wireless Metropolitan Area Network (WMAN) technology, especially in developing countries. NS-2 does not have an official WiMAX package, but the user community, including organizations such as NIST [113] and the WiMAX Forum [99], as well as independent researchers [108] have contributed NS-2 models that implement the IEEE 802.16 standard to varying degrees. This example focuses on the NIST IEEE 802.16 model since it is open source and available to all NS-2 users. Table 3-9 compares the PHY features of the NIST model with other WiMAX models from QualNET and OPNET. Although the NIST model implements some IEEE 802.16-2005 features, it does not implement the OFDMA PHY or FEC [113].

TABLE 3-10. IEEE 802.16 Frame Duration PHY Example Parameters

Baseline	Comparison
Mac/802_16 set frame_ duration_ 0.005	Mac/802_16 set frame_ duration_ 0.010

The purpose of this example is to again reinforce the importance of accurately matching system parameters that are model inputs to the real-world scenario being simulated. If the actual parameters are unavailable, multiple points that test the boundary conditions should be used as inputs to identify any output anomalies. The authors adjusted the example simulation script "datarate.tcl" that is part of the NIST IEEE 802.16 model distribution to investigate the average application throughput of a single mobile subscriber station (SS) communicating with a single BTS. In this example, the frame duration is set to 5 ms and 10 ms and the CP is set to the commonly implemented IEEE 802.16-2004 values of 1/4 and 1/16. Both frame durations and CPs are simulated with the BPSK and QPSK burst profiles, which are generally used for control traffic. The burst profile and CP are specified at the NS-2 command line when executing the modified "datarate.tcl" script. The frame duration is set in the Tcl script by modifying the frame duration parameter as shown in Table 3-10.

A standard NS-2 trace file is produced, in this case using the "new" wireless trace format that contains more information than the "old" format produced in the IEEE 802.11 examples [111]. A Perl script is used to determine the average data rate for each combination of frame duration, modulation type, and CP. Simulation results are plotted in Figure 3-7. The combinations of modulation type and CPs plotted in the left bars correspond to a frame duration of 10 ms and those on the right correspond to a frame duration of 5 ms. As illustrated in Figure 3-7, doubling the frame duration allows a larger payload to be carried on the downlink and the system less frequently alternates between DL and UL. Therefore, the average data rate is increased for all modulation types and CPs simulated. As a sanity check, Figure 3-7 also shows that QPSK achieves a higher average data rate than BPSK, since twice as much data are carried in a QPSK symbol than in a BPSK symbol.

This WiMAX example illustrates a typical PHY simulation comparison using NS-2. Other recommended simulation exercises for the interested reader are to compare the results in Figure 3-7 with a frame duration of 20 ms. Additional SSs could also be added to the scenario and the impact on average UL data rate from a particular SS could be investigated. The impact of the DL ratio parameter on system performance is also of interest, including the minimum and maximum DL ratio allowed in the simulation.

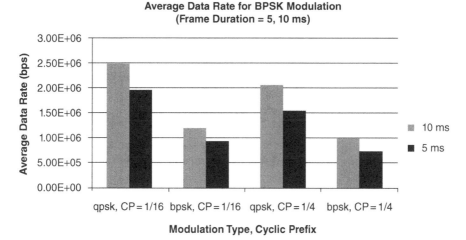

FIGURE 3-7. Average data rate for simulated combinations of modulation type, CP, and frame duration.

3.4 WIRELESS NETWORK SIMULATION LESSONS LEARNED AND COMMON PITFALLS—PHY LAYER

The main topics of wireless network PHY simulation have been presented in this chapter, with specific detail given to the subtopics of 1) incorporating interference into a PHY model, 2) the importance of considering the PHY preamble, and 3) practical PHY NS-2 model examples. The focus of these subsections was to describe possible consequences of PHY oversimplification in a wireless network simulation and to provide scenarios where the omission of details may or may not be acceptable. The astute reader may have inferred that the motivation for discussing these topics in detail comes from the fact that many simulation designers frequently omit these details in their simulations. Multiple sources including [1, 4, 35, 36, 50, 51], of which the authors are aware, describe the areas in which wireless network simulations are lacking details that can make published results questionable. The intention of this section is to describe some of these common simulation pitfalls as they apply specifically to the PHY in hopes that the reader may apply best practices to improve future simulations.

One of the most widely cited papers that motivates PHY simulation best practices is [52], which compares simulation results based on a set of common axioms found in Mobile Ad Hoc Network (MANET) academic research papers. The six axioms simplify the wireless environment to be modeled and therefore the job of the simulation designer, but at the cost of severely impacting the application of results to a given environment. The six axioms follow [52]:

1. The world is flat (i.e., there are no signal obstacles such as hills, trees, buildings, etc.).
2. A radio's transmission area is circular (i.e., all radios use ideal omni-directional antennas).
3. All radios have equal transmission range (i.e., all antennas have equal gain, all radios have equal power output).
4. All radio links are symmetrical (i.e., if node A can hear node B, then node B can hear node A).
5. The propagation environment is perfect (i.e., no fading loss and no inter-ference or channel contention when transmitter is within range of the desired receiver).
6. Signal strength is a simple function of distance (i.e., only separation distance between nodes impacts signal strength, not obstructions or scattering).

The six axioms are not necessarily inappropriate for all simulation scenarios, but are generally very simplistic. The simulation designer must decide which metrics he or she desires as the simulation output and whether or not making any of the assumptions would corrupt those results; however, in any case where the wireless PHY or MAC is to be evaluated, the designer should be careful incorporating any of these axioms. Additionally, when simplification through applying these axioms is chosen, this information should be conveyed with any presented results from the simulation. A detailed discussion of the six axioms can be found in [52] to help the designer avoid these common simulation pitfalls.

A list of best practices in PHY simulation to help the simulation designer avoid common pitfalls is summarized:

- Use caution when assuming that multi-user interference is Gaussian. If this assumption is true, published results should include evidence to enforce this assumption
- The PHY preamble must be considered as special bits and not simply as additional random data bits when higher layer protocols are being evaluated
- Choose an appropriate radio model for the simulation scenario. If sup-ported by the simulator, include standard terrain and mobility models and provide those data to the research community with any published results in standard formats
- Choose the target environment being simulated carefully and verify that the simulator parameters match this environment
- A statistically valid number of trials must be executed in the simulation. Researchers should consider calculating confidence intervals and adding to any published results

- The model developer should choose a simulation platform with which he or she is most familiar

Most simulation results are collected using a single simulator because the learning curve for any given simulator, whether it is commercial, open source, or a homebrew solution is extremely steep [4]. The problem that the developer faces is not getting an output, but getting the correct output that is applicable to the desired scenario. Default parameters in a simulation may not capture the scenario of interest and include embedded parameters that must be fully understood and reported along with results. For example, the particular methods by which a simulator calculates interference and noise contributions must be understood if the results are to be meaningful. As illustrated in Table 3-1, common wireless network simulators handle interference calculations differently.

The researcher who presents simulation results should be careful to specify all conditions under which the results were obtained. This includes the number of nodes, geographic distribution, mobility model considered, transmission range, fading models, and whether the communication was unidirectional or bidirectional. The inclusion of this information not only allows others to independently validate results, but it allows the research community to slowly begin to put more confidence into wireless network simulation results. If sufficient space is not available in the research article to allow these details to be included, external links can be embedded as a reference where additional simulation settings can be found. Additionally, simulation source code could also be shared with the research community to jump-start other researchers in verifying results and extending work to the next level [41].

CHAPTER 4

Medium Access Control Modeling and Simulation

While the primary focus of this book is wireless network simulation, many of the common simulators used by the research community such as NS-2 and OPNET began as wired network simulators and have been extended to the wireless environment. With this in mind, also consider that most practical networks are heterogeneous mixtures of wired and wireless sub-networks. Therefore, this chapter provides the simulation designer with an overview of common wired and wireless network MAC-layer protocols and associated principles that should be incorporated into MAC-layer simulations.

Much of this chapter focuses on the question of how much detail in a MAC-layer simulation is sufficient. That is, when comparing a fully detailed simulation of a MAC protocol such as CSMA/CD or CSMA/CA to a very abstract, bare bones implementation, what is the cost in terms of accuracy of results? As the reader will learn, this mapping between an abstract simulation and a detailed simulation depends on the particular scenario being modeled. The simulation designer is always faced with a dilemma of how much detail to include in a simulation. The simulation designer must also choose the correct level of cross-layer interaction in a network simulator. For example, a simulation that considers the PHY at the bit-level or packet-level may not necessarily provide more accurate results than a flow-level simulation. Again, an acceptable amount of layer interaction depends on the particular scenario to be simulated and the simulator to be used.

In many cases simulation is the only available option for characterizing the behavior of MAC protocols since analytical models describing CSMA/CD or CSMA/CA mechanisms become intractable when all behavioral details are included. Network simulation is also generally preferred over experiments that require testbed equipment due to the cost and availability of test equipment and often cumbersome experimentation setup and configuration.

An Introduction to Network Modeling and Simulation for the Practicing Engineer, First Edition.
Jack Burbank, William Kasch, Jon Ward.

Simulation allows scalability considerations that are not possible in real-world testbeds with a limited number of nodes. This chapter presents some of the fundamental wired and wireless MAC concepts to be considered in simulation design. The MAC models supported by the common simulators NS-2, OPNET, GloMoSim, and QualNET are also presented. Two NS-2 wireless MAC simulation examples are included to further reinforce the concepts described. This chapter concludes with suggestions to the reader by reviewing some of the lessons learned and common pitfalls of wired and wireless MAC network simulation.

4.1   MODELING AND SIMULATION OF WIRED MACs

While the focus of this book is wireless networks, this section discusses the topic of modeling and simulation of MAC layers from wired technologies, such as Ethernet and Token Ring, for completeness.

4.1.1   The Token Ring LAN

Token ring is a wired LAN data link layer technology that uses a round-robin method of token passing to designate which station has the privilege to transmit [24]. Each node is connected to the ring with two cables, one with which it receives data from its upstream neighbor and another which is used to send data to a downstream neighbor. A single token is passed throughout the network and may be in the possession of only one node at a time. The node that currently has possession of the token may transmit data to other nodes on the network; nodes without the token must refrain from transmitting until they receive the token. A timer is also associated with the token such that a single node may not seize the channel indefinitely. Once a node has finished transmitting or the timer has expired, the token is passed to the downstream neighbor. The token ring network is included here for completeness and has been largely replaced by Ethernet as the preferred LAN medium [25, 30]. Token ring LANs are still frequently found in the financial industry at banks and at retailers. In these applications, token ring meets the throughput demands of the users and has therefore not been replaced. OPNET offers token ring simulation capability [15]; the capability of NS2 and QualNET to model token ring is unknown since token ring simulation is not an active research area [13, 26].

4.1.2   The Ethernet LAN with CSMA/CD

The Ethernet CSMA/CD LAN MAC protocol mimics the way a room full of people talk to one another. There is no central controlling entity in the network and any node may exchange frames with another node. All nodes must first perform carrier sensing on the copper or fiber cable to determine whether the channel is occupied. If carrier sensing determines that the medium is currently in use, the node defers until the medium becomes available. Once the medium is available, the message may be placed on the line. If multiple transmitters attempt to transmit simultaneously, they end transmission as soon as they become aware of the other nodes' transmissions, transmit a jamming signal to notify all nodes on the LAN that a collision has occurred, and enter an exponential back-off period. This period sets a random timer for each node that determines how long a node must delay attempting to retransmit [15]. Once the exponential back-off timer has expired, the node again returns to the carrier sensing state and assesses the availability of the channel. If the carrier is active, the node refrains from transmitting; otherwise, it seizes the channel and transmits its frame. During the transmission the node continues to listen to the channel since the length of the LAN may induce delays that could cause transmissions to collide, as illustrated in Figure 4-1.

The continual monitoring of the link while transmitting is referred to as collision detection and is unique to wired networks. Because all nodes in the LAN are connected to a common wired medium, an active transmitter is able to determine that a collision has occurred if another transmitter's message is heard on the medium. In the example from Figure 4-1, Node 1 first senses that the channel is idle and begins its transmission. Because there is a propagation delay between Nodes 2 and 1, in the second scene Node 2 also senses the

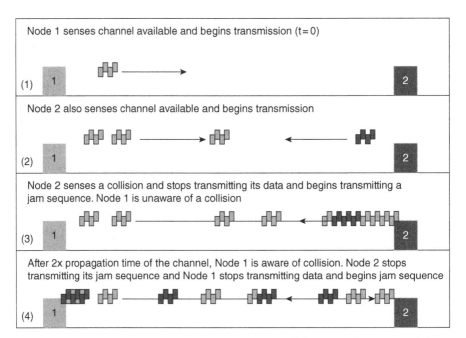

FIGURE 4-1. Illustration of example CSMA/CD collision scenario, recreated from [27].

channel as idle since Node 1's transmission has not yet arrived at Node 2 due to the propagation delay of the channel. In scene 3, Node 2 becomes aware of Node 1's transmission through its collision detection mechanism of listening to the channel during its transmission. At this point, Node 2 ends its data transmission, transmits a 32-bit jam sequence for Ethernet channels up to Gbps, and enters random back-off [35]. At this point Node 1 is still unaware of the collision since it continues to listen to the channel, but the propagation delay is such that Node 2's original transmission and jam signal have not yet reached Node 1. In scene 4, after twice the propagation time of the channel, Node 1 receives Node 2's transmission and infers that a collision occurred that corrupted its original transmission. Node 1 ends its transmission, transmits the 32-bit jam signal, and enters random back-off [27]. Collision detection is a luxury available in wired networks, but not in wireless networks; wireless MAC protocols such as CSMA/CA must handle the unreliable wireless channel with no guaranteed feedback from the distant receiver to indicate that a collision occurred.

4.1.3 How Much Detail Is Required in a Wired Ethernet Simulation?

The details of two independent studies comparing the impact of details on simulation results for the Ethernet CSMA/CD protocol are described in this section as a case study [28, 29]. As expected, the conclusion from both of these studies is that abstraction is a useful tool to reduce simulation complexity and to speed up debugging and execution time, but results are only applicable in certain scenarios. The simulation designer must also be careful to understand all results and to determine any impact that abstraction may have on those results. As previously described, the CSMA/CD protocol has multiple components that require significant signaling overhead in a simulation implementation. A fully detailed CSMA/CD implementation must have the following components: propagation delay on the LAN, exponential back-off, carrier sensing, and a jamming signal mechanism. While these four CSMA/CD components do not seem large, the designer must also consider the scalability of the simulation and the fact that these four components must interact over the N nodes present in the network. The larger question posed at the beginning of this chapter now becomes which of these four components are required to be implemented in a CSMA/CD implementation to produce accurate simulation results? The simulation designer must consider how much error can be tolerated in the result. If the simplification saves the designer four weeks of implementation time and the simulation executes five times faster than the fully detailed model, but the results over-estimate throughput by 20%, is this within the acceptable margin of error? Of course the answer to this question is scenario dependent and largely depends on the end application of the results.

Both [28] and [29] examine the effects of abstract implementations of CSMA/CD, given that these models have become commonplace in the wired network simulation community. The lack of research validating CSMA/CD simulations is likely due to the fact that most wired simulations, much like current wireless simulations, focus on validation of upper-layer protocols such as routing protocols and applications. Researchers have tended to neglect the effects of the MAC and PHY on simulation results, a trend that will need to be addressed as the community becomes more interested in cross-layer network designs. Historically, network researchers have modeled wide area links as a queue with fixed propagation delay and constant bit rate and Ethernet links have been modeled as a simple point-to-point link. The question then is are these simplified models truly acceptable or are they widely used because network researchers have little interest in the lower layers of the protocol stack? Furthermore, if these simplified models are determined to accurately represent CSMA/CD, over what range of conditions are they valid?

In [29], an abstract model of CSMA/CD is implemented using a drop tail queue, which simplifies the shared access protocol and contention mechanism. In this model, all nodes on the network transmit data frames to a common queue which then transmits the data frames onto the shared medium on a first in, first out (FIFO) schedule. The drop tail queue only places a single node's frames on the channel at one time, hence removing the need for carrier sensing or a collision mechanism since collisions cannot occur. A single queuing delay is added to the model to account for the exponential back-off condition. Frame loss, as in the IEEE 802.3 CSMA/CD specification, occurs after a retransmission fails 16 times and the retransmitted frame is discarded. The drop tail queue models packet loss only when the queue becomes full and other packets waiting to be queued are dropped [29]. The experiment in [29] compares a real-world testbed to an IEEE 802.3 CSMA/CD simulation in NS-2 and the drop tail queuing model, also implemented in NS-2. The results demonstrate that the NS-2 simulation closely matched the actual testbed results by predicting a more conservative 7% lower network throughput. The simplistic drop tail model predicted a consistently higher throughput of 22% greater than the CSMA/CD simulation and the testbed. The simple drop tail model also reduces simulation run time by 75% compared with the NS-2 CSMA/CD simulation. Armed with this background information, the simulation designer must then choose whether or not the particular simulation application can tolerate a potential 22% error in results to achieve a 75% increase in execution time as well as the time savings in implementing the simple simulation compared with the fully detailed simulation.

In [28], three versions of the CSMA/CD protocol are implemented in simulation with varying levels of detail and each are run under equivalent traffic conditions to determine the impact of abstraction on the accuracy of results. Furthermore, the three levels of simulation detail provided in [28] allow the effects of abstraction to be observed and compared with the fully detailed CSMA/CD model with finer resolution than [29]. The three models considered are labeled as the completely abstract model, partially abstract, and fully detailed. The completely abstract model ignores propagation delay and assumes all nodes are immediately aware of any other node transmitting. The partially abstract model takes the abstract model and implements the exponential back-off from [30]. The fully detailed model implements the propagation delay on the LAN link and exponential back off. These three models are tested in a scenario with 30 clients and a single server. For an average network load of less than 40%, all models achieve equivalent throughput predictions due to the low number of collisions and retransmissions. Beyond a 40% average load, the throughput predictions diverge. This is compared with the average run time, which grows slowly for the abstract and partially abstract models under increasing average load conditions, but grows exponentially for the fully detailed model. The results in [28] achieve a similar conclusion as the results for [29]. For a moderate average network load of less than 40%, the simplified models produce results that are acceptable and the performance savings is substantial. But care must be taken for high average loads of 70% and above, where the results for the simplified models are significantly different than the fully detailed model. In fact, for these high average load scenarios, the simplified models may overestimate the achievable throughput of the LAN by 50% [28].

These examples demonstrate the importance of characterizing a model over a wide range of variable parameters. It also demonstrates that in some cases an abstract model is acceptable and in other cases the abstract model does not produce accurate results. The simulation designer must select the appropriate level of detail for the particular experiments being conducted and the metrics being collected [28].

4.1.4   Wired MAC Simulator Implementations

The commonly used network simulators NS-2, OPNET, and QualNET each offer wired MAC modeling capabilities (see Table 4-1). OPNET appears to have the most comprehensive wired MAC support. The interested reader is referred to references [13, 15, 26] for more detailed information on MAC models supported by each of these simulators.

TABLE 4-1. Wired Network Simulation Models Available in NS-2, OPNET, and QualNET

NS-2 (v. 2.1b8)	OPNET	QualNET
Ethernet (CSMA/CD) [13]	Fiber Distributed Data Interface (FDDI), Gigabit Ethernet (and others), Token Ring [15]	IEEE 802.3 (CSMA/CD), Gigabit Ethernet, Wired Point-to-Point Link MAC protocol [16]

4.2 WIRELESS NETWORK MAC SIMULATION

In a wireless network, the MAC has the primary function of moderating access to the shared wireless medium by defining rules that allow multiple wireless devices to communicate with one another in an orderly and efficient manner [34]. Much research has been done on wireless MAC design because it exemplifies the traditional engineering tradeoff, and the solution must be optimized to the specific application under consideration. Typical metrics used to evaluate MAC performance are throughput, delay, stability, contention resolution, fairness, and power savings. The classic research problem for wireless MAC design has been to develop a MAC that maximizes throughput and fairness while minimizing delay. Recent research has considered quality of service (QoS) and bandwidth-intensive multimedia services such as Internet Protocol television (IPTV) or the less strenuous case of voice over IP (VoIP). In this case, the MAC design must provide some priority to users that require more immediate service than other users with less time-sensitive applications (e.g., email, web browsing, or file transfer). In the sensor network field, researchers have shifted attention to wireless MACs that maximize power savings to allow sensors to maximize their lifetime in the field [39, 40].

The intention of this section is to describe some of the basic building blocks of wireless MAC design and topics that should be considered by the wireless simulation designer. This section will also present some of the most frequently referenced MAC designs that are implemented in the four common wireless network simulators described in this chapter: GloMoSim, NS-2, OPNET, and QualNET. As can be seen from Figure 4-2, recreated from [41], this collection of four simulators composes approximately 25% of all simulators from which published results are generated. In fact, this number is most likely much larger, but the "others" category that composes 50% of the graph incorporates papers that did not specify the simulator.

MAC Layer Simulators used in Surveyed Wireless Network Papers

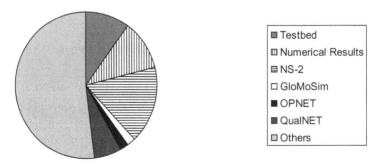

FIGURE 4-2. Most commonly used MAC layer simulators [41].

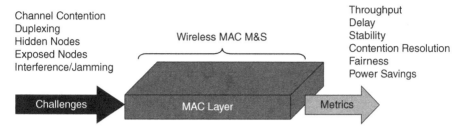

FIGURE 4-3. Wireless network MAC simulation example.

Whether the simulation designer is implementing a new MAC protocol or using one already implemented in a wireless network simulator, there are basic characteristics of a MAC protocol that should be understood. Some of these characteristics are illustrated in Figure 4-3.

The MAC protocol must have a duplexing mechanism that allows it to multiplex transmitted and received traffic. For example, IEEE 802.11 is a half duplex, asynchronous, time division duplex (TDD) system. An IEEE 802.11 radio is not capable of receiving and transmitting at the same time, which makes it half duplex and the transmitter and receiver are not synchronous from frame to frame. The same frequency is used by both the transmitter and receiver, making it a TDD system. A system where the transmitter and receiver are frequency separated such as the UL and DL of most cellular telephone systems are frequency division duplexed (FDD). These systems are often full duplex because the receiver is capable of receiving and transmitting at the same time since the signals are spectrally separated.

Unlike wired MAC protocols, the transmitter is generally unaware that a collision occurred. For this reason, wireless MAC protocols attempt to minimize the probability of a collision at the desired receiver. Wireless MAC protocols may be binned into two broad categories, depending on the type of network for which they are designed: distributed and centralized MAC protocols. Figure 4-4 shows this classification of MAC protocols in more detail.

Wireless MAC protocols may be further binned as random access, guaranteed access, and hybrid access protocols. In random access protocols, nodes contend for access to the medium. ALOHA [42] is the classic example of a random access protocol. In a guaranteed access protocol, nodes access the medium in an orderly manner, usually according to a round-robin schedule. An example of a guaranteed access protocol is a polling protocol where a master polls the slave for channel access or a token passing protocol, where only the node in possession of the token may access the channel. Hybrid access protocols are a blend of random and guaranteed channel access and are usually based on a request and grant relationship [34]. In a hybrid access MAC, a client node may request necessary bandwidth from a central node and the central node would then allocate an upstream time slot for the data to be transmitted. This is similar to the IEEE 802.16 MAC, where a subscriber

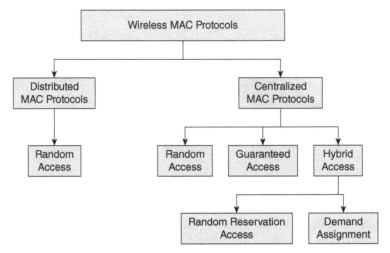

FIGURE 4-4. Classification of wireless MAC protocols, recreated from [34].

station may reserve an uplink time slot with a bandwidth request MAC message [37]. Hybrid access protocols may be further binned as random reservation (RRA) and demand access protocols. In a RRA system, a node may request access to the uplink channel and the central node decides whether or not to grant the request based on a set of predefined rules. In a demand assignment protocol, a client may request access to the uplink channel and the central node makes a decision based on achieving its current QoS requirements [34].

Because the transmissions of a wireless medium are not restricted to a cable, wireless MAC protocols must allow nodes to communicate reliably in unfavorable RF conditions. Two situations that may arise in a wireless network, depending on node geometry, are the hidden node and exposed node problems. Both are briefly explained in subsequent paragraphs. Channel contention and interference from other users competing for the channel must be handled in a stable manner. The wireless MAC protocols such as CSMA/CA, IEEE 802.11, and Multiple Access with Collision Avoidance (MACA) have generally been tested thoroughly by researchers under various conditions; however, simulation designers that are attempting to implement new wireless MAC protocols must consider the various tradeoffs and the wide range of conditions over which the MAC must operate reliably.

4.2.1 The Hidden Node Problem

In a wireless network, not every node can communicate with all other nodes due to physical obstructions or exceedingly great distances between the MSs. The inability of each MS to hear one another creates a situation referred to

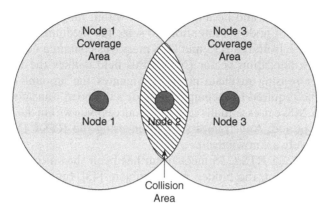

FIGURE 4-5. Example of hidden node problem in a generic wireless network.

as the hidden node problem, which is depicted in Figure 4-5. The hidden node problem is one of the most difficult challenges for the wireless MAC designer to overcome while still being fair and maintaining application throughput.

In Figure 4-5, both nodes 1 and 3 can communicate with node 2. Nodes 1 and 3, however, cannot communicate with one another because of range and possible obstacles. Therefore, node 1 is unable to determine whether node 3 is transmitting because node 1 does not receive any frames transmitted from node 3. The same holds for node 3 since it cannot determine whether node 1 is transmitting. This prevents both nodes 1 and 3 from performing either physical carrier sensing or virtual carrier sensing; however, node 2 receives transmitted frames from both nodes 1 and 3. As shown in Figure 4-5, this situation causes a collision area because neither node 1 or 3 will know to back off from the other node's transmission when communicating with node 2. The hidden node problem is quite common and generally difficult to detect in systems such as IEEE 802.11 because of the half-duplex nature of transceivers. That is, since IEEE 802.11 transceivers transmit and receive on the same frequency, MSs are unable to make clear channel assessments (CCAs) while simultaneously transmitting.

To combat the hidden node problem, the IEEE 802.11 standard allows stations to use request to send (RTS) and clear to send (CTS) messages to effectively allow all MSs to be aware of transmissions in a given area. The transmitting MS initially transmits a RTS frame to the MS with which it wishes to communicate. All MSs within range of the transmitting MS are able to receive and process the RTS frame. The intended recipient then responds by sending a CTS frame and all MSs within range receive and process the CTS. In the example shown in Figure 4-5, node 1 wishes to exchange data with node 2. Node 1 first sends a RTS to node 2 and all nodes inside of the left circle receive the RTS and learn node 1's intentions of seizing the wireless medium. Node 3 is out of range to receive node 1's messages to node 2. Node 2 responds with a CTS to node 1 and because node 2 is within range of nodes 1 and 3, both nodes learn of node 1's intentions to seize the medium for communications with node 2. Both the RTS and CTS messages contain a field referred to as the network allocation vector (NAV); this field defines the IEEE 802.11 virtual carrier sensing mechanism and designates the amount of time the channel will be required to accomplish a node's intended transmission. Using this approach, MSs can essentially reserve resources even within the contention-based access period. Any frames sent when using the RTS/CTS procedure must be positively acknowledged.

The IEEE 802.11 RTS/CTS mechanism has been shown to mitigate, to at least a certain extent, the hidden node problem [43]. Literature shows that when RTS/CTS procedures are employed, the performance of the IEEE 802.11 distributed coordination function (DCF) is only marginally dependent on the minimum contention window size and the number of active MSs (i.e., RTS/CTS is relatively scalable) [44]. However, not surprisingly, this RTS/CTS approach is relatively inefficient from a capacity usage perspective because of the additional latency introduced before the transmission can begin. That is, a tradeoff exists between decreased network throughput due to the additional RTS/CTS overhead versus the increased reliability of avoiding IEEE 802.11 frame collisions from other contending MSs. As a result, RTS/CTS is only employed when the size of the packet to be transmitted exceeds a parameter called the RTS threshold. Otherwise, the standard DCF channel access procedures are employed. RTS threshold is often configurable through the IEEE 802.11 device driver, but the authors have rarely seen RTS/CTS used in practice. This is likely due to the fact that most IEEE 802.11 networks operate in infrastructure mode and the DCF is sufficient for mitigating interference due to contention. It should be noted that the hidden node problem as described in this section assumes that nodes operate with omni-directional antennas. If nodes 1 and 2 exchange RTS and CTS messages using directional antennas, the handshaking messages do not reach node 3 and have no impact on the scenario. This point is made to reinforce the fact that the simulation designer must completely understand the scenario to be modeled before he or she can understand the model output.

4.2.2   The Exposed Node Problem

The exposed node problem is similar to the hidden node problem in that it decreases network throughput due to interference. An exposed node is defined as a node that is within the range of the sender but out of the range of the destination. In systems employing a CSMA MAC protocol, all nodes must refrain from transmitting once their carrier sensing mechanism reports that the channel is busy. Leveraging the illustration in Figure 4-5, assume that node 2 attempts to transmit a frame to node 1. Node 3 detects node 2's transmission through its carrier sensing mechanism and refrains from transmitting while node 2 transmits its frame to node 1. The problem here is that in the geometry of this example, node 1 cannot hear node 3 such that node 3's transmissions would not cause interference at node 1. By adhering to the CSMA protocol, node 3 wastes the opportunity to transmit by waiting until the channel is unoccupied, yet nodes 2 and 3 could simultaneously transmit without interfering with the recipient of the respective frames. In a network with many nodes such as a dense field of sensors or perhaps a battlefield where soldiers carry distributed sensors, the network throughput decreases significantly as many nodes become subject to the exposed node problem and refrain from transmitting during opportunities that would not cause interference [34]. The exposed node problem assumes that all users transmit with approximately the same output power; that is, in our example, if node 3 is significantly more powerful than node 2, it will most likely cause interference at Node 1.

4.2.3   The Impact of Interference on the Wireless MAC

Interference may have many impacts on a wireless MAC that should be considered by the simulation designer. If the MAC is not designed properly, a wireless network may suffer from the near-far problem, where a loud interferer that is closer to the desired receiver than the desired transmitter is more likely to have its message received because its signal is overwhelmingly larger than the desired transmitter's signal [34]. A properly designed MAC must implement fairness by ensuring that multiple nodes' transmissions are offset from one another and do not arrive simultaneously at a given receiver. The IEEE 802.11 MAC uses the randomized back-off to enforce asynchronous transmission in the system and prevent this condition of interference. The wireless MAC should also be able to operate properly in conditions of extreme interference such as that experienced when a wireless network is maliciously jammed.

4.2.4   The Multiple Access Collision Avoidance MAC Protocol

In CSMA protocols, a node that has data to send first senses the channel to determine whether the channel is idle before transmitting. If the channel is busy, the node waits for a random period of time and, once its timer expires, attempts to access the channel again. If the node senses that the channel is idle, the node attempts to acquire the channel and transmit data. If a collision occurs when the node attempts to access the channel, the colliding nodes resolve the collision according to the specific MAC protocol [34, 39]. Under the assumption that significant collisions occur due to the hidden node problem, the MACA protocol uses a three-way handshake of RTS/CTS/Data. This is much like the IEEE 802.11 MAC that uses a four-way handshake of RTS/CTS/Data/Acknowledgment (ACK), where an ACK is returned by the receiver after the data frame is received correctly [34, 39].

4.2.5 The IEEE 802.11 CSMA/CA MAC Protocol

CSMA/CA is provided by the DCF, which is the predominant contention mechanism used in IEEE 802.11 networks. It should be noted that IEEE 802.11 does not employ CSMA/collision detection (CD) as used by the IEEE 802.3 LAN standard because collisions are considered too costly due to the limited amount of bandwidth present in IEEE 802.11 networks. The IEEE 802.11 DCF allows multiple independent MSs to interact without central management. Like Ethernet, the DCF first checks to ensure that the wireless medium is available before transmitting a frame. To avoid collisions, MSs use a random back-off time after each transmitted frame, where the first MS to transmit seizes the channel. Two types of carrier sensing functions are employed by IEEE 802.11 to determine whether the medium is available: physical carrier sensing and virtual carrier sensing. If either of these functions determines that the medium is not available, the MAC reports this unavailability to the higher layers.

An important mechanism in coordinating access to the medium is interframe spacing. The IEEE 802.11 MAC employs four different interframe spaces: short interframe space (SIFS), point coordination function (PCF) interframe space (PIFS), DCF interframe space (DIFS), and extended interframe space (EIFS). These varying interframe spacings provide the ability for IEEE 802.11 to support different priority levels for different types of traffic. This is illustrated in Figure 4-6.

As can be seen from Figure 4-6, the interframe spacings are used in conjunction with the CSMA contention protocol previously outlined. Upon observation of an idle medium, the MS waits some prescribed period of time before contending for channel access. That period of time is defined by the interframe spacing. SIFS is used for highest-priority transmissions because its length in time is the shortest of all interframe spacings, allowing a MS with a high-priority frame to seize the channel before any other MSs have the opportunity to contend. Therefore, the high-priority control traffic frames transmitted in the SIFS are transmitted before the longer PIFS, DIFS, or EIFS intervals. Frames transmitted using SIFS include RTS/CTS frames and positive ACKs. PIFS is used by MSs operating under the PCF to gain priority access to the

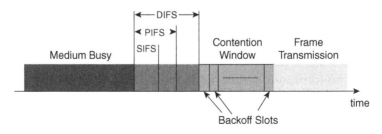

FIGURE 4-6. IEEE 802.11 DCF channel access [46].

medium for contention-free operation; MSs that have data to transmit in the contention-free period can transmit after the PIFS has elapsed and effectively preempt contention-based traffic. The PIFS is 30 μs for the IEEE 802.11b PHY, although according to [46], no commercial products implement PCF. DIFS is the minimum medium idle time for contention-based services; the DIFS is 50 μs for IEEE 802.11b. MSs operating in the contention-based mode may have access to the medium if it has been free for a period longer than the DIFS. This is the normal data traffic mode of operation for IEEE 802.11 WLANs. The EIFS, which is not shown in Figure 4-6, is not a fixed interval and is only used when a CRC error is detected in the received frame [46].

As shown in Figure 4-6, there is a period called the contention window or back-off window that follows the interframe spacing. This window is divided into time slots, where the slot length is PHY-dependent and is smaller for higher data rate PHYs. MSs pick a random slot and wait for that slot before attempting to access the medium. All slots are equally likely. As in Ethernet, the back-off time (i.e., number of slots) is selected from a larger range each time a collision occurs. Contention windows are always sized 2^{n-1}, for $n \geq 1$. For each collision that occurs, n is incremented by 1. Each PHY has a maximum back-off time; for example, the original IEEE 802.11 DSSS PHY has a maximum back-off time of 1023 time slots. The IEEE 802.11b PHY has a minimum contention window size of 31 and the maximum contention window size is 1023. The corresponding slot time for the IEEE 802.11b PHY is 20 μs. The contention window size increases exponentially within the constraints set by the maximum and minimum window size. Correspondingly, this also increases the amount of time that is required to gain access to the channel to transmit data, increasing the latency of the system as seen by those data.

Once the MS has waited the selected back-off period of time, the MS will determine whether the medium is in use through physical and virtual carrier sensing (i.e., CCA). If the medium is not idle, the MS defers access and waits for the medium to become idle again. If the medium is still idle after the interframe spacing, the MS can begin its random backoff timer and transmission can begin in the chosen slot. If the previous transmitted frame was received without error, the medium must be free of transmissions for at least the interframe spacing. If the previous transmission contained an error, the medium must be free for the amount of the EIFS to allow for retransmission. The DCF medium access procedure is depicted in Figure 4-7 [17].

4.2.6 The IEEE 802.16 (WiMAX) MAC Protocol

The previous sections of this chapter have focused on CSMA and specifically the IEEE 802.11 MAC. The reason for this is because these MAC protocols are widely available in network simulators, including GloMoSim, NS-2, OPNET, and QualNET. The intention of this section is to expose the reader to an alternate wireless MAC design that is quite different from the IEEE 802.11 MAC, the IEEE 802.16-2004, or fixed WiMAX MAC [37]. This section

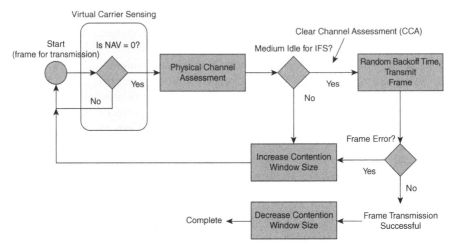

FIGURE 4-7. DCF channel access functional diagram.

includes only an overview of the IEEE 802.16 MAC and is by no means comprehensive. The interested reader is referred to the IEEE 802.16-2004 standard for additional information [37]. Another outstanding source of information for both IEEE 802.11 and on IEEE 802.16-2004 is [48].

The IEEE 802.16 MAC comprises three parts: the Service Specific Convergence Sub-layer (CS), The MAC Common Part Sub-layer (MAC CPS), and the Security Sub-layer, as shown in Figure 4-8.

This section briefly describes the functions of each sub-layer. The CS is intended to handle two types of predominant traffic: Asynchronous Transfer Mode (ATM) and IP traffic. The majority of IEEE 802.16 networks transport IP traffic. The CS sub-layer exists between the higher layers of the protocol stack and the MAC CPS. Its purpose is to classify traffic since all traffic in IEEE 802.16 networks is associated with a connection. Connections, or service flows as they are called in IEEE 802.16 terminology, are then mapped to Service Flow Identifiers (SFIDs) and Channel Identifiers (CIDs). The MAC CPS handles the traditional MAC functions as described in this section including duplexing, channelization and message structuring, and initial network access. The IEEE 802.16 networks allow both TDD and FDD duplexing as specified in [37, 48], but the majority of actual deployed systems are TDD. Channelization is achieved through time division multiplexing (TDM), where frames are created and transmitted in appropriate time slots. This creates a fundamental difference in the IEEE 802.16 MAC compared with the IEEE 802.11; the IEEE 802.16 MAC is necessarily synchronous and only exists in an infrastructure deployment. The BTS controls all aspects of channel access and provides the SSs with uplink bandwidth on a frame-by-frame basis. All SSs must negotiate with the BTS for initial network access. The security sub-layer provides privacy through encryption of the link between the BTS and

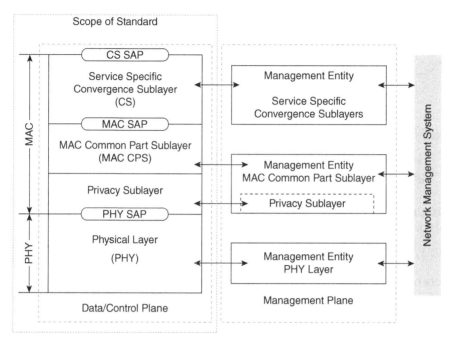

FIGURE 4-8. IEEE 802.16-2004 logical architecture that demonstrates the three MAC sub-layers [37].

SS. Network authentication and encryption of select MAC messages and traffic is accomplished through various mechanisms such as X.509 digital certificates, Secure Hash Algorithm (SHA-1), Triple Data Encryption Standard (3DES), and Advanced Encryption Standard (AES).

4.2.7 Wireless MAC Simulator Implementations

GloMoSim, NS-2, OPNET, and QualNET each offer a variety of wireless MAC protocols from which the simulation designer can choose. This section describes the particular MAC protocol implementations for each of the four common simulators. There are most likely additional MAC protocols available for each of the simulators that are not discussed in this section, especially custom add-ons for NS-2 and GloMoSim. This section is intended as a starting point for the interested reader and not as an exhaustive catalog of each simulator's capabilities.

NS-2 offers the simulation designer many options for wireless MAC implementations. Prior to NS-2 version 2.33, there was only a single IEEE 802.11 MAC provided by Carnegie Mellon University (CMU) that implemented the Distributed Coordination Function (DCF) for an ad hoc IEEE 802.11 network. There was previously no NS-2 MAC implementation for an infrastructure

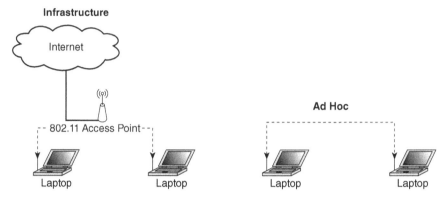

FIGURE 4-9. Wireless network topology example: infrastructure network (left) and ad hoc network (*right*) [17].

network [13]. These two IEEE 802.11 network architectures are depicted in Figure 4-9.

NS-2 versions beyond 2.33 offered multiple options for IEEE 802.11 network simulation. The older ad hoc implementation was still available as well as an extension that implements IEEE 802.11 infrastructure networks that includes device scanning, authentication, and association.

Besides IEEE 802.11, NS-2 implements unslotted ALOHA [42], a basic MAC with no collision detection for a single receiver (MAC/sat), and a time division multiple access (TDMA) MAC protocol. The TDMA protocol is single hop and preamble-based, where every wireless node has a dedicated transmission slot. Other MAC protocols exist for NS-2 such as the IEEE 802.16 MAC [49, 99, 113], which are supported by the NS-2 user community and not included in the official NS-2 package. In the spirit of open source programming, NS-2 also offers hooks for the simulation designer to implement his or her own MAC protocol. Some of these functions include, but are not limited to [13]:

- txstop() – Returns the time when the channel will become idle, which can be used by the MAC to implement carrier sense.
- contention() – Allows the MAC to contend for the channel before sending a packet. The channel then uses this packet to signal the corresponding MAC object at the end of each contention period.
- collision() – Indicates whether a collision occurs during the contention period.
- send() – Allows the MAC object to transmit a packet on the channel for a specified duration of time.
- hold() – Allows the MAC object to hold the channel for a specified duration of time without actually transmitting any packets.

GloMoSim also offers the simulation designer multiple MAC protocols that are already implemented. These include CSMA, MACA, and IEEE 802.11 [14]. The GloMoSim version of CSMA includes carrier sensing, but does not implement CA to mitigate the hidden node and exposed node problems. Therefore, throughput is potentially degraded when using this MAC protocol due to these two scenarios in the wireless environment being modeled. The GloMoSim MACA protocol adds a RTS/CTS exchange to the CSMA protocol to mitigate the hidden node and exposed node problems. The MACA and IEEE 802.11 MAC protocols differ in the requirement for an acknowledgment message by the receiver after successful reception of a data frame. The IEEE 802.11 implementation includes this ACK to allow the transmitter to immediately retransmit a lost or corrupt frame when no ACK is received from the desired receiver [14]. GloMoSim also allows the simulation designer to implement MAC extensions if the three wireless MAC protocols do not meet his or her needs. The reader is referred to [14] for more details on creating GloMoSim MAC extensions.

Both OPNET and QualNET implement a large number of MAC protocols for both wired and wireless channels. The reader is referred to [15] for a complete list of OPNET MAC protocols and [16] for a complete list of QualNET MAC protocols. A subset of the OPNET and QualNET wireless MAC protocols are included here for fair comparison with the capabilities of NS-2 and GloMoSim. OPNET implements CSMA, CSMA/CA, and slotted ALOHA, where slotted ALOHA is provided by the user community [15]. QualNET implements ALOHA, MACA, TDMA, CSMA, the IEEE 802.11 PCF, the IEEE 802.11e Enhanced Distributed Channel Access (EDCA), and the IEEE 802.11e Hybrid Coordination Function Controlled Channel Access MAC Protocol.

Table 4-2 lists the wireless MAC protocols present in specific versions of GloMoSim, NS-2, OPNET, and QualNET. As can be seen, all support an IEEE 802.11 implementation, which justifies its frequent use in MANET simulations. Table 4-2 is not intended to be exhaustive and the interested reader is referred to [13–16] for a more detailed description of each simulator's capabilities.

TABLE 4-2. MAC Models Available in GloMoSim, NS-2, OPNET, and QualNET

GloMoSim (v.2.02)	NS-2 (v. 2.33)	OPNET	QualNET
CSMA, MACA, TSMA, IEEE 802.11 [14]	IEEE 802.11, CSMA/CA, MAC/sat, Unslotted ALOHA, TDMA [13]	CSMA, CSMA/ CA, Slotted ALOHA* [15]	IEEE 802.11 PCF, IEEE 802.11e EDCA & HCCA, ALOHA, CSMA, GMAC, MACA, TDMA [16]

*Denotes this model and support as being provided by user community.

4.3 PRACTICAL MAC MODEL IMPLEMENTATIONS

The examples in Section 3.3 motivated the need for carefully choosing PHY parameters that apply to the real-world scenario being simulated. This point applies to choosing MAC parameters as well. This section includes two NS-2 simulation examples that vary MAC parameters that have previously been described in this chapter. 'Again, the objective here is not to demonstrate whether or not the NS-2 IEEE 802.11 MAC implementation represents any given wireless scenario, but rather to illustrate that the NS-2 output is completely dependent on the MAC parameters chosen by the simulation designer. Section 4.3.1 illustrates an IEEE 802.11b example where the application-layer throughput varies with the defined RTS threshold. Section 4.3.2 illustrates an IEEE 802.16-2004 example where an appropriate queue length is chosen such that no MAC PDUs are dropped.

4.3.1 IEEE 802.11b RTS Threshold MAC Example with NS-2

As described in Section 4.2.1, IEEE 802.11 networks may use RTS to mitigate the hidden node problem. Although not usually used in practice, RTS may be configured by way of the RTS threshold parameter. The RTS threshold defines a frame size where any frames equal to or exceeding the threshold shall be preceded by a RTS message. Therefore, setting the RTS threshold high (e.g., 3000 B) means that no frames will use RTS since this number exceeds the Ethernet maximum transmission unit (MTU). Correspondingly, setting the RTS Threshold small (e.g., 30 B) will require most frames to use RTS. In this example, two IEEE 802.11b two-node cases are simulated, where in one case the RTS threshold is set to 3000 B and in the other case it is set to 30 B. The purpose of this example is to demonstrate that the overhead associated with using RTS has a considerable effect on the application-layer throughput of the simulated system. Table 4-3 illustrates the changes made to the "simple-wireless.tcl" simulation description in this example [109, 112].

Table 4-4 summarizes a subset of possible output parameters processed by applying a Perl script to the NS-2 output trace file, where the RTS threshold set to 3000 Mbps case is shown on top and the RTS threshold set to 30 case on the bottom. When using the RTS threshold set to 3000 B, the application-layer throughput was approximately 36% more than the case where the RTS threshold is set to 30 B (e.g., 2.385978 Mbps compared with 1.538197 Mbps).

TABLE 4-3. IEEE 802.11b RTS Threshold MAC Example Parameters

Baseline	Comparison
Mac/802_11 set RTSThreshold_ 3000	Mac/802_11 set RTSThreshold_ 30

TABLE 4-4. IEEE 802.11b RTS Threshold MAC Example Output

```
====== Analyzing NS-2 Wireless Data (RTS Threshold =
  3000 B)======
Data Packets Sent           : 20985 pkts (11333400 B)
Data Packets Received       : 20953 pkts (11742220 B)
Simulation Duration         : 146.544735678 s
Avg. Application Throughput  : 2385978.94736842 (bps)


====== Analyzing NS-2 Wireless Data (RTS Threshold =
  30 B)======
Data Packets Sent           : 13511 pkts (7306440 B)
Data Packets Received       : 13479 pkts (7547780 B)
Simulation Duration         : 148.031916585 s
Avg. Application Throughput  : 1538197.89473684 (bps)
```

TABLE 4-5. Summary of WiMAX Features Implemented in Common Network Simulators Recreated from [108]

Feature	QualNET [97]	OPNET [98]	NS-2 NIST [113]
Standard	IEEE 802.16e	IEEE 802.16e	IEEE 802.16e
Duplex Mode	TDD	TDD	TDD
Hybrid ARQ	No	No	No
Multi-Hop	No	No	No
ARQ	Yes	Yes	No
QoS	Yes	Yes	No
Handoff	Yes	Yes	Yes

In this example, the 144 bit long IEEE 802.11b preamble is used for both simulation cases, with a data rate of 11 Mbps. The RTS threshold values of 3000 B and 30 B were purposely chosen to be extreme conditions; however, they illustrate the importance of matching the RTS threshold parameter with the actual IEEE 802.11b equipment settings being used in the real-world scenario if throughput results are to be considered valid.

4.3.2 NS-2 WiMAX MAC Example (IEEE 802.16 NIST Model)

The NIST NS-2 WiMAX model is introduced in Section 3.3.2 in a simulation that examines PHY parameters. In general, the NIST implementation offers much more flexibility to the simulation designer at the PHY than at the MAC via the Tcl script. Although the designer is always able to modify the underlying C++ functions of a given NS-2 library, this is not for the NS-2 novice. Table 4-5 compares the MAC features of the NIST model with other WiMAX models from QualNET and OPNET [108, 113].

TABLE 4-6. IEEE 802.16 MAC Queue Example Parameters

Baseline	Comparison
`Mac/802_16 set queue_` `length_ 500`	`Mac/802_16 set queue_length_ x` x ranges from 50 to 20,000 B

One MAC parameter that is available for configuration through the NS-2 Tcl script is the size of the queue for received messages. In this example, the authors demonstrate how NS-2 could be used to determine an acceptable size for an ingress queue that does not drop any MAC PDUs. In general, a more realistic problem statement would be to define an acceptable percentage of packets that may be dropped since there will always be scenarios that exceed the incoming queue size; however, for purposes of illustration, our objective is to determine an appropriate queue size that drops zero MAC PDUs over the duration of the simulation.

The authors adjusted the example simulation script "datarate.tcl," as in Section 3.3.2, and the scenario considers a single SS communicating with a single BTS. In this example, the frame duration is set to 5 ms, and the modulation type and CP are specified as BPSK and 1/4, respectively. The queue length is varied from 50 B to 20 kB in the Tcl script by modifying the queue_length_ parameter as shown in Table 4-6. The simulation duration is specified as 50 s.

The results are parsed using a Perl script and the number of dropped MAC PDUs is plotted versus the chosen queue size in Figure 4-10. The results illustrate a linear relationship between the number of dropped MAC PDUs and the size of the ingress queue. For this simulated scenario, a queue size of approximately 19.750 kB achieves zero dropped MAC PDUs. Perhaps dropping 25 % of the total amount of frames that would be dropped with no incoming queue is acceptable to the real-world application. That is, with a zero-length queue, approximately 20,000 frames are dropped. Based on the simulated results in Figure 4-10, choosing a queue size of approximately 14.5 kB would drop approximately 5,000 frames on average. The purpose of this example is to illustrate one of the biggest advantages of simulation, the ability to quickly analyze the output based on a wide range of inputs. Once an acceptable queue size is chosen via simulation, it is always recommended that simulation results be compared to results from hardware-in-the-loop (HITL) or other real-world experiments.

4.4   NETWORK SIMULATION LESSONS LEARNED AND COMMON PITFALLS—MAC LAYER

This chapter presents Ethernet, IEEE 802.11, and IEEE 802.16 MAC designs and some of the parameter tradeoffs associated with each. It should be clear

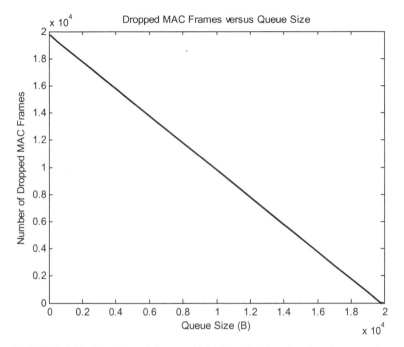

FIGURE 4-10. Number of dropped MAC PDUs for simulated queue size.

to the simulation designer that there is no "one size fits all" MAC to cover all wired and wireless network architectures. For the simulation designer who is focused on modeling the behavior and performance metrics of higher-layer protocols, the MAC protocols in Table 4-2 are most likely sufficient. For the simulation designer who desires to implement and test a new wireless MAC protocol, some of the most widely reported wireless MAC performance metrics are included here [34, 39]:

- Delay – The average time spent by a frame in the MAC PDU ingress queue from the moment it is queued until the transmission is complete
- Throughput – The fraction of the channel's capacity that is used for MAC frame transmission
- QoS – Delay-intolerant applications such as IPTV, VoIP, or streaming radio must be provided to the user in a timely manner. The wireless MAC protocol must refrain from queuing frames associated with these applications for an excessive amount of time
- Fairness – A MAC protocol must not exhibit preference to any single node when multiple nodes are attempting to access the wireless medium

- Stability – The overhead associated with the MAC protocol may be significant in terms of headers and trailers attached to the higher-layer payload. The MAC must remain stable under loading conditions that approach the capacity of the channel without exhibiting erratic behavior
- Robustness to Interference – The MAC must remain stable under interference conditions such as a deep fade, jamming, or a surge in users competing for the channel
- Power consumption – This is the key metric for wireless sensor networks. Since most consumer wireless devices have limited battery power, MAC protocols must implement power-saving features to conserve battery life

There are inherent tradeoffs in these performance metrics that the MAC designer must address. Throughput is a conflicting goal with delay; the optimal solution for a wireless MAC design maximizes throughput while minimizing queuing delay. Evaluation of throughput and delay performance must also consider the MAC overhead associated with headers, trailers, control, and management traffic and not be simply based on upper-layer payload data. Fairness is also a conflicting goal with QoS, since giving priority to a given node to support QoS for its application forces the MAC design to be biased [34, 39]. Power consumption is always a conflicting goal because the more complexity that is added to an algorithm, the more processing and ultimately more power consumption required. Certainly, implementing power-savings algorithms such as those used in IEEE 802.11, Bluetooth, ZigBee, and cellular networks increases delay and may decrease throughput. Other performance metrics are bounded and simulation may be the only method of determining exactly what those bounds are. For example, the stability of the MAC protocol is only demonstrated by testing the offered network load from zero to full capacity to determine the breaking points. Similarly, the robustness to interference objective is bounded by the duration and magnitude of interference. Important data points such as the maximum duration of fade tolerable or the minimum SINR before the link becomes unusable are most easily found through simulation. Simulation may also be used for evaluating future systems where DSA considerations and cross-layer interaction between the PHY, MAC, and upper layers is necessary for networking algorithms to function appropriately.

Wired network simulation has been studied by researchers since the 1970s and many conventions have been established in this field. Yet research questions remain concerning the level of detail required in wired MAC simulations to produce accurate results. Generally, the level of detail required in a simulation depends on the specific problem at hand and must be estimated by the simulation designer. In this case, the designer must understand the problem at hand sufficiently to include enough detail to make the simulation results meaningful. Much of this chapter focuses on the amount of abstraction that can be tolerated in a specific simulation. Abstraction is an important tool in simulation because it can offer the following benefits:

- Savings in run time of the simulation
- Simplification in simulation implementation
- Decreased simulation debugging time

Of these three benefits, only the savings in run time can be easily quantified. The simulation run time for the fully detailed simulation may be compared with the abstract simulation run time, and the difference in execution time is considered a savings. The simplification in simulation implementation and the decreased simulation debugging time are difficult to quantify because they are subjective metrics that may be valued differently by each simulation designer.

Some of the common pitfalls of wired and wireless simulation are described in the following paragraphs. First, the simulation designer should collect all known parameters relating to the actual system being modeled and incorporate them into the model as early as possible to increase the likelihood that simulated results accurately reflect the behavior of the actual system. Additionally, the simulation designer should determine how much error in the results could be tolerated in the specific application. Some applications have large error margins while others require extremely accurate results. Second, in making comparisons such as those made in this chapter between simplified models, abstract models, and fully detailed models, the same higher layer traffic must be provided in all simulation scenarios to make a fair comparison. This is commonly chosen as constant bit rate (CBR) traffic in published literature. Even beyond the traffic models, all conditions must be held constant across the experiments to achieve the most meaningful comparisons.

Third, all simulations should be tested with conditions that stress the system and find bounds where results may no longer be valid. This is seen in [28], where varying the offered load on the network over a wide range of values demonstrates regions where the simplified model does not match the fully detailed model. Equivalently, researchers should not draw trends based on a limited amount of data points that do not consider simulation behavior at boundary conditions. Lastly, model abstractions must be careful not to exclude interactions with other layers of the network stack. For example, the tail drop queue abstraction used in [29] as a substitute for CSMA/CA does not consider any PHY interaction. This may or may not be acceptable. For a scenario that considers the maximum throughput of an Ethernet link with an offered load of 30%, the PHY interaction is probably not very important. If the simulation designer must characterize an optimal length of twisted pair category-5 cable for minimizing retransmissions at a switch on an Ethernet link, the effects of the PHY should be included.

In general, the simulation designer should not lose sight of the fact that a simulation is an abstraction to the real world and, as such, often gives the designer control to change parameters that may not be available in the actual equipment or scenario being modeled. As shown in the examples in Sections 4.3.1 and 4.3.2, changing the value of relatively simple parameters could have a dramatic effect on the trends of output metrics. This is why it is so important that all known parameters relating to the actual system being modeled be collected and incorporated into the model as early as possible to increase the likelihood that simulated results capture the essence of the actual system.

CHAPTER 5

Modeling and Simulation for Higher Layer Protocols

Up to this point, our discussion has focused on the lower layers of the protocol stack. And as can be seen, successfully modeling the PHY, MAC, and RF propagation environment are in themselves challenging tasks. However, whether you are modeling a wired or wireless network it is important to adequately represent the protocols that will be operating over the network. This section provides an overview of the capabilities of various simulation platforms regarding support of higher layers of the protocol stack.

5.1 NETWORK LAYER

The network layer is arguably the most difficult layer of the protocol stack to model in a meaningful way due to the sheer scope of the problem. At the PHY and MAC layer, an extreme amount of detail is sometimes required for meaningful simulation (i.e., bit-level simulation). Transport and application layers also may require a significant amount of detail to provide meaningful results; however, to include significant detail in a model or simulation of the network layer is often times impractical. What if the researcher is interested in understanding the behavior of a routing protocol in a large network of hundreds of nodes, thousands of nodes, or even tens of thousands of nodes? What if the Internet itself is the system to be modeled? The sheer size of this problem space mandates abstraction. This problem is exacerbated if a wireless environment is to be considered. Fluctuating RF channel conditions and mobility-induced topology changes can severely affect the performance of a network's routing system. Thus, these effects cannot be ignored. But how are they implemented in a large-scale simulation? There are also practical considerations in the modeling of the network layer such as memory requirements. How does

An Introduction to Network Modeling and Simulation for the Practicing Engineer, First Edition.
Jack Burbank, William Kasch, Jon Ward.
© 2011 Institute of Electrical and Electronics Engineers. Published 2011 by John Wiley & Sons, Inc.

one model a large network and the behavior of a routing protocol without requiring a heroic amount of memory to execute the simulation?

It is critical to understand the importance of abstraction at the network layer. If you are a network researcher implementing a new protocol in simulation, you must decide how much detail is appropriate to capture the desired behavior. A highly detailed simulation is likely going to result in a simulation that is too slow to yield meaningful results for a large-scale network in a timely manner; however, an overly simplified model will also not yield meaningful results. If you are utilizing predeveloped protocol implementations found in a commercial or open-source simulation tool, it is important to note that this abstraction is likely already present at the network layer. The designers of the particular protocol have already gone through the process of deciding the proper level of detail *in their determination* and have implemented simplified algorithms. This is important to realize for two reasons: *1) two network simulation tools may yield vastly different results for the exact same simulation scenario and protocol, and 2) a particular network simulation may be designed to capture behaviors different than the behaviors you want to model.* When using preexisting protocol implementations, it is imperative that you understand what has been implemented to ensure that it models the behavior you are interested in and to the appropriate degree of detail to meet your particular needs. Methods are suggested in [53] to deal with varying levels of detail, including error randomization and visualization techniques that can help isolate incorrect details and deal with too much detail.

It is important to also note that at some point, it is not only impractical but likely impossible to model a large network. A compelling argument is made in [54] that simulating the Internet is not possible, if for no other reason that it is not well enough understood in real life. What is the real topology of the Internet? What are the traffic flows? Even if all the processing power was available to support a high-fidelity model of the Internet as a whole, [54] argues that the Internet is not well enough understood to model. This book argues that this is true in general. At some point, systems grow in complexity to the point that they cannot be well enough understood, even by their designers, to be modeled perfectly.

Even if a large network is well understood, there are practical considerations. Let us return to the example of the Internet. Making some conservative assumptions about the Internet, [55] shows the staggering processing and memory requirements likely required to model a system the size of the Internet. In the example constructed in [55], approximately 1.4 petabytes of disk storage would be required to log the results from a 100 second simulation of the Internet and that this would require 30 billion CPU seconds to simulate. While the exact numbers are a function of simulation implementation, this does illustrate the difficulty in simulating large networks. In [56], another example is given of the computing complexity of simulating network layer protocols. In the example of [56], approximately 5 gigabytes of memory is required to simulate a multicast routing protocol. This single protocol alone exceeds the capacity of many modern computing systems. There are two primary methods in which implementers attempt to overcome the problem of complexity: 1) parallelism and distributed computing, and 2) simplification. Chapter 7 discusses distributed methods in more detail. Work conducted by researchers in the area of routing protocol simulation have shown that dramatic improvements are possible through compression, aggregation, and the removal of superfluous routing states [57–59]. The work in [59] reduced the memory requirements of the multicast routing protocol to an insignificant amount compared to the total simulation requirements; however, even [59] noted that future work would be to migrate their simulation into a parallel environment to enable even larger simulations. This suggests that even with simplification, parallel computing is a key enabler for large-scale simulation.

5.1.1 Network Layer Protocol Support in Existing Simulation Tools

Table 5-1 summarizes the network layer protocol support in some of the main network simulation tools. Tables 5-2 and 5-3 go on to summarize the network

TABLE 5-1. Network Layer Protocols in GloMoSim, NS-2, QualNet, and OPNET

QualNet	GloMoSim (v.2.02)	NS-2 (v. 2.1b8)	OPNET
IPv4, IPv6, Dual-IP Stack, Mobile IPv4, HSRP	IPv4	IP, Mobile IPv4, Mobile IPv6, NEMO	IPv4, IPv6, Mobile IPv4, Mobile IPv6, HSRP, RSVP

TABLE 5-2. Wired Network Layer Routing Protocols in GloMoSim, NS-2, QualNet, and OPNET

QualNet	GloMoSim (v.2.02)	NS-2 (v. 2.1b8)	OPNET
Static, RIP, BGPv4, BGPv6, EIGRP, IGRP, OSPFv2, OSPFv3	N/A	Static, RIP, OSPF, BGPv4, IS-IS	Static, RIP, BGP, EIGRP, IGRP, IS-IS, OSPF, OSPFv3

TABLE 5-3. Wireless Network Layer Routing Protocols in GloMoSim, NS-2, QualNet, and OPNET

QualNet	GloMoSim (v.2.02)	NS-2 (v. 2.1b8)	OPNET
AODV, BRP, DSR, DYMO, Fisheye, IARP, IERP, OLSR, OLSRv2, STAR, ZRP	AODV, DSR, Fisheye, ODMRP, WRP	AODV, OLSR, DSR, Fisheye, DYMO	AODV, DSR, GRP, OLSR, TORA

layer routing protocols that are currently supported in these simulation tools. These lists are by no means exhaustive.

Some observations can be made at this point:

- Generally speaking, GloMoSim supports a smaller set of network layer protocols as compared with its counterparts.
- OPNET and QualNet, both commercial tools, support a rich set of network layer protocols for both wired and wireless environments.
- QualNet has extensive support for wireless network layer protocols, supporting all major adhoc routing protocols under development to date.
- Commercially available tools generally have a far greater set of supported protocols as compared with open source counterparts.
- It is sometimes difficult to even determine what is available for opensource tools since they are typically user-created and not always centrally located. Even when an implementation of a particular protocol is available, it sometimes requires effort to obtain.

5.1.2 Mobility Models

As mentioned in Chapter 1, mobility models can have a significant impact on the performance of a network. This is most pronounced at the MAC layer and network layer and the network's routing system. Unfortunately, mobility models are typically arbitrary. It is typically difficult to characterize the mobility of current users much less predict the mobility of future network users. In this sense, wired network simulation is much easier than wireless network simulation. In the wired network mobility is not a consideration. But in the wireless network, mobility cannot be ignored. The results presented in [60] demonstrate that different mobility models can yield completely different performance results for the same adhoc routing protocol. Figure 5-1 presents results from [60] for a mobile network employing the Destination-Sequenced Distance Vector (DSDV) proactive adhoc routing protocol. Here we can see that while the general performance trends are largely the same (i.e., degrading performance with increasing speed), the actual results can vary dramatically at any one particular node speed value (e.g., 10 m/s produces widely varying results). This result demonstrates the sensitivity of simulation results to mobility assumptions.

It is important to model mobility adequately if performance statements are going to be made about a routing protocol. Yet these models are arbitrary. What this means is different depending on the purpose of the simulation. If the simulation is being used to assess the performance of a particular existing network deployment, then great care should be taken in understanding the mobility of the users of the network and sensitivity to network performance should be understood. If a network researcher is investigating the general

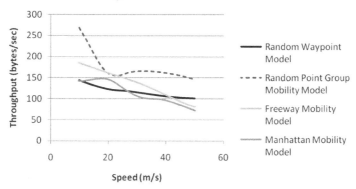

FIGURE 5-1. An example of the impact of mobility on simulated protocol performance.

performance of a particular protocol, then mobility models should be chosen such that a fair comparison can be made between protocols and across research papers in open literature.

The most commonly employed type of mobility model is a random way-point (RW) mobility model [61] where a node's location (i.e., destination) and velocity are randomly chosen (typically) according to a uniform distribution defined by a minimum and maximum velocity for each discrete simulation instance. Another commonly used mobility model, particularly in the military community, is the Random Point Group Mobility (RPGM) model [60]. Here, nodes are grouped into clusters where a cluster is defined by some area about a center point. Within the cluster nodes are typically randomly distributed. The cluster as a whole moves according to a RW-type model to provide macro-level mobility. There can also be intra-cluster mobility where individual nodes are allowed to move within the cluster boundary, typically also accord-ing to a RW-type model, providing micro-level mobility. Another mobility model often employed, both in the commercial and military domain, is a freeway mobility (FM) model. In the FM model nodes are constrained to move in straight lines, notionally representing movement by vehicles on a road. Nodes can either be placed equidistant from each other or some ran-domness employed for inter-vehicle spacing (typically according to some uniform distribution). Nodes can then either move at identical velocities (i.e., their relative distance never changes) or can have velocity fluctuations so that the inter-vehicle distance changes over time. There are many, many other mobility models—too many to mention here in a reasonable manner. The reader is referred to [61] and [62] for a discussion of mobility models for use in ad-hoc networks. The reader is also referred to [63] and [64] for a discussion of the IMPORTANT framework, which is a tool for analyzing mobility in NS-2-based simulations.

When utilizing existing mobility models already integrated into existing simulation tools, the reader is urged to fully understand the details of these mobility models so that informed performance statements can be made.

Another example of the impact of mobility on network-layer performance can be found in the work presented in [117], which compares several network-layer mobile adhoc network (MANET) routing protocol statistics for various mobility models. Five primary mobility models are considered in [117]: random direction, Manhattan, city section, Gauss-Markov, and random waypoint. The paper presents results from their home-brew Java simulation environment showing the sensitivity of routing protocol performance to the assumed mobility model. In [117], under Random Waypoint mobility conditions, the average hop count between mobile nodes varies from two to three. However, if a random direction mobility model is assumed, the average hop count ranges from three to four.

Additional examples of the impact of mobility on network layer routing protocol performance in simulations can be found in [118–120].

5.2   TRANSPORT AND APPLICATION LAYERS

Table 5-4 summarizes the transport layer protocol support in some of the main simulation tools.

It is clear from Table 5-4 that the commercial simulation tools provide a rich set of transport layer protocols. There is also a surprisingly large amount of transport layer protocols available for NS-2. There is basic support in GloMoSim for TCP and UDP.

Table 5-5 lists some of the applications supported by these simulation tools. Again, this list is not meant to be exhaustive; however, it is clear from Table 5-5 that the commercially available tools provide a rich set of application layer protocols.

TABLE 5-4. Transport Layer Protocols in GloMoSim, NS-2, QualNet, and OPNET

QualNet	GloMoSim (v.2.02)	NS-2 (v. 2.1b8)	OPNET
TCP, TCP Lite, TCP Reno, TCP New Reno, TCP SACK, TCP Tahoe, UDP, RTP	TCP, UDP	TCP, TCP Reno, TCP New Reno, TCP SACK, TCP Tahoe, TCP Westwood, RTP	TCP, TCP ECN, TCP Reno, TCP New Reno, TCP SACK, TCP Tahoe, UDP, PEP, SCPS-TP, RTP

TABLE 5-5. Application Layer Protocols in GloMoSim, NS-2, QualNet, and OPNET

QualNet	GloMoSim (v.2.02)	NS-2 (v. 2.1b8)	OPNET
CBR, VBR, FTP, HTTP, Ping, SNMP, Telnet, Generic traffic generators	Telnet, FTP	CBR, VBR, VoIP, FTP, HTTP	CVR, VBR, VoIP, Video, Telnet, HTTP, FTP, Email, Database, Generic traffic generators

5.2.1 Traffic Models

As discussed in Chapter 1, almost any performance statement made regarding a network must be conditioned on the traffic loading of that network. Therefore, it is important to accurately reflect the traffic offered to the network; however, traffic models are often arbitrary and are almost always guaranteed to not be precisely correct. It is important to perform sensitivity analysis to understand the impact of discrepancies in simulated versus real traffic on the performance of the network. If using existing traffic generation tools, either in open source or commercial packages, it is important to understand exactly how those tools are generating traffic. Additionally, it is possible that even if using an open source or commercially available simulation package that the simulation designer might need to develop representations of the applications that utilize the simulated network. There are a plethora of traffic generation tools available to the simulation implementer. The reader is referred to [65–67], just to name a few, for some tools that can be used to generate traffic within the NS-2 simulation environment. Many additional traffic generation tools exist if one wishes to simulate traffic in a real hardware environment (e.g., [68]). Table 5-6 summarizes some of the network traffic generation tools currently available. Note that Table 5-6 is by no means exhaustive. It should also be noted that the testbed-related tools listed in Table 5-6 are likely more useful when considering a hardware-in-the-loop approach, which will be discussed in Chapter 6.

Historically, the most commonly employed traffic model was the Poisson model. The Poisson model, however, does not exhibit the type of burstiness exhibited by computer network traffic. Consequently, variations of the Poisson model, such as the Markov-modulated Poisson model [121] and the compound Poisson traffic model [122] were proposed to introduce some degree of data burstiness. In [123] it was demonstrated that Ethernet computer networks exhibit self-similarity, where self-similarity refers to distributions that look similar regardless of the time-scale under consideration, which is not the case for Poisson models. This finding of self-similarity was further confirmed by

TABLE 5-6. Network Traffic Generation Tools

Traffic Generation Tool	Notes	URL
NSWEB	WWW Traffic Generator for NS-2	http://www.net.t-labs.tu-berlin.de/~joerg/
PackMime-HTTP	HTTP Traffic Generator for NS-2	http://www.dirt.cs.unc.edu/packmime/
TMix	TCP Workload Generator for NS-2	http://ccr.sigcomm.org/ online/?q=node/50
Harpoon	Traffic generator for testbeds	http://pages.cs.wisc.edu/~jsommers/ harpoon/
D-ITG	Internet traffic generator for testbeds	http://www.grid.unina.it/software/ITG/
GenSyn	Internet traffic generator for testbeds	http://www.item.ntnu.no/people/ personalpages/fac/poulh/gensyn
Ixia*	Traffic generator for testbeds	http://www.ixiacom.com/
iPerf	TCP and UDP traffic generator for testbeds	http://sourceforge.net/projects/iperf/

*Indicates commercial product.

[124]. This means that there are some networks for which a Poisson representation is not appropriate. Another important finding was presented in [125], which concluded that TCP congestion control does not cause self-similar network traffic, nor does it mitigate this property. The authors of [125] go on to show that self-similar traffic models are the best choice if there is large-scale aggregation of network traffic, and they also state that if the sample set is extremely small, then traditional Poisson models are usually sufficiently accurate. Ultimately, it is up to the simulation developer to understand the type of network being simulated so that the proper choice in traffic model is made. If the developer is utilizing home-brew traffic sources, it is important to properly understand the nature of the network to be simulated and its source traffic. Even if off-the-shelf traffic generation tools are to be employed, it is important to understand how they work to decide whether they are adequate or not for the network to be simulated.

5.3 EXAMPLE OF HIGHER LAYER MODELING: TRANSPORT LAYER PERFORMANCE ANALYSIS

Many of the available simulation tools allow for various degrees of abstraction to achieve targeted analysis of particular layers of the protocol stack. As an example, consider Figure 5-2, which depicts a simple simulation setup in OPNET.

In this simulation, the IP cloud abstractly simulates a network of IP routers, with the major attributes of packet latency (PL) and packet discard ratio (PDR). In this simple simulation, TCP is employed with attributes listed in Table 5-7.

The application is configured to be a large FTP data file transfer. This type of simple setup allows for targeted analysis of TCP protocol performance in various network conditions without the need for detailed network layer simulation. A sample result from this simple simulation is shown in Figure 5-3. This figure shows decreasing throughput performance as the error rate of the abstracted network increases. This type of simulation allows the developer to gain insight into transport or application layer protocols over a generic network with generalized performance attributes. This type of abstracted network layer simulation can be useful in observing performance trends of transport or application layer protocols or in fine tuning of transport or application layer protocol parameters.

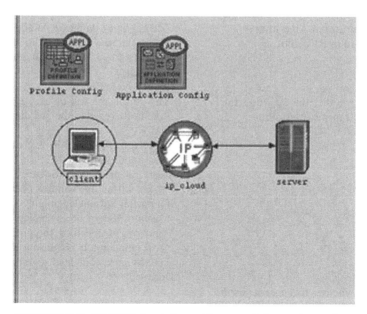

FIGURE 5-2. OPNET simulation with network layer abstraction.

TABLE 5-7. TCP Attributes Used in OPNET TCP Simulation Example

Attribute Name and Description	Simulation Values
Maximum Segment Size (MSS) (bytes) Largest data size that will be transmitted as one TCP segment	8152 / 1460
Receive Buffer Size (bytes) Size of the buffer holding received data before it is forwarded to application layer	8760 / 32768 / 65536
Receive Buffer Threshold Threshold used to determine the limit on the usage of receive buffer before transferring data out to the socket buffer	0
Maximum ACK Delay (sec) Maximum time the TCP waits after receiving a TCP segment before sending an ACK	0
Slow-Start Initial Count (MSS) Specifies the number of MSS-sized TCP segments that will be sent upon slow start. Also specifies the value of the initial congestion window	1 / 2 / 4 / As defined in RFC 2414* RFC 2414 upper bounds this initial window as Min[4*MSS, max(2*MSS, 4380 bytes)]
Fast Retransmit Indicates whether this host uses the Fast Retransmit Algorithm described in RFC 2001	Enabled* *TCP will retransmit segments on detection of duplicate ACKs. Congestion window will be in slow start (set to 1 MSS initially) after recovery.
Fast Recovery Indicates whether this host uses the Fast Retransmit Algorithm described in RFC 2001	Reno* / New Reno** *Fast retransmit will be executed once the node receives the 3rd duplicate ACK. **Fast retransmit will be executed with two modifications: fast retransmit will never be executed twice within one window of data. If a partial acknowledgment (acknowledgment advancing snd_una) is received, the process will immediately retransmit the next unacknowledged segment.
Window Scaling Indicates whether this host sends the Window Scaling enabled option in its SYN message	Disabled

TABLE 5-7. Continued

Attribute Name and Description	Simulation Values
Explicit Congestion Notification (ECN) Capability Specifies if TCP implementation supports ECN. Both sides must exchange support for ECN before making use of this feature	Disabled
Selective ACK (SACK) Indicates whether this host sends the Selective Acknowledgment Permitted option in its SYN message	Disabled / Enabled
Segment Send Threshold Determines the segment size, and granularity of calculation of slow start threshold variable	Byte Boundary* *During fast retransmission, slow start threshold will be set to half of the current congestion window.
Nagle's Algorithm Enabling this algorithm prevents small segments from being sent while the sender is waiting for data acknowledgment	Disabled
Karn's Algorithm Enabling will cause the Karn's algorithm to calculate the retransmission timeout values	Enabled
Timestamp Specifies whether TCP timestamp option (RFC 1323) is supported. The TCP timestamp option provides better estimates of the TCP round-trip time calculation. It also protects against wrapped sequence numbers	Disabled
Initial Sequence Number Initial segment sequence number used for all connections from this node	0
Retransmission Threshold Specifies the criteria used to limit the time for which retransmission of a TCP segment is done	Maximum connect attempts = 3 Maximum data attempts = infinite
Initial Retransmission Timeout (RTO) (sec) Initial RTO value used before the RTO update algorithm comes into effect	3

TABLE 5-7. Continued

Attribute Name and Description	Simulation Values
Minimum RTO (sec)	1
Maximum RTO (sec)	64
Round Trip Time (RTT) Gain	0.125
Gain used in updating the RTT measurement	
Deviation Gain	0.25
Gain used to update the mean round trip deviation	
RTT Deviation Coefficient	4
Coefficient used to determine the effect of mean deviation on the final calculated RTO value	
Timer Granularity (sec)	0.5
Represents TCP slow timer duration (used to handle all timers except maximum ACK delay timer). Timer events are scheduled at multiples of this value	
Persistence Timeout (sec)	1
Duration of the persistence timeout. This allows the local socket to receive a window update when the receiver window is very small	

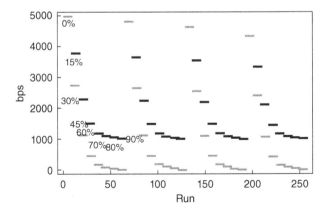

FIGURE 5-3. TCP performance for various packet error rates.

5.4 EXAMPLE OF HIGHER LAYER MODELING: DETAILED NETWORK LAYER MODELING

The various network simulation tools also allow for detailed modeling of the network layer itself to gain insight into the performance of the network. Figures 5-4 and 5-5 show a detailed network topology in the OPNET simulation environment. Figure 5-4 depicts a group of mobile platforms, with Figure 5-5 showing the network topology within each mobile platform. Figure 5-6 shows the detailed network diagram for the intra-platform network of Figure 5-5, which shows a mixture of wired infrastructure and wireless connectivity based on IEEE 802.11.

Tables 5-8 and 5-9 summarize the attributes of this detailed Ethernet/IEEE 802.11 simulation. The goal of this particular example simulation was to investigate the relationship between network performance and IEEE 802.11 transmit power, with the ultimate goal of learning how low this power can be tuned down while maintaining acceptable performance.

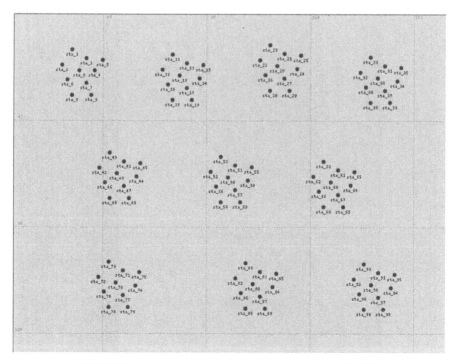

FIGURE 5-4. Network topology modeled within OPNET.

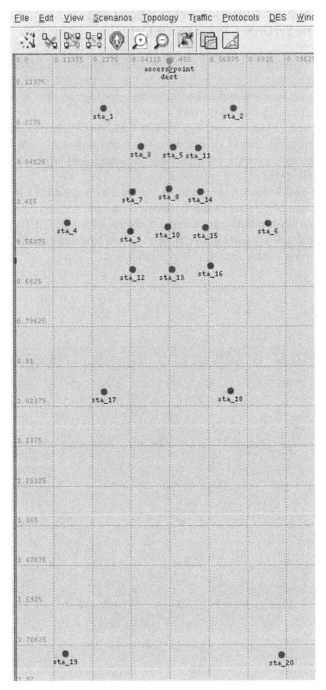

FIGURE 5-5. Intra-platform network topology modeled within OPNET.

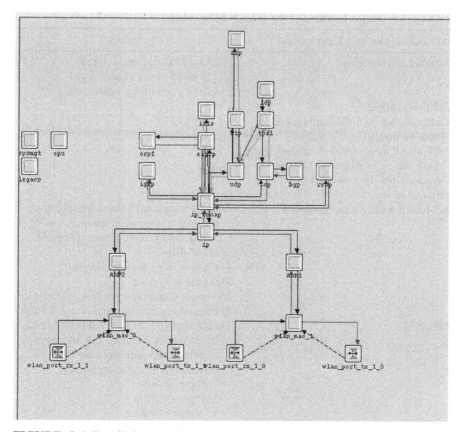

FIGURE 5-6. Detailed network diagram with both Ethernet and IEEE 802.11 connectivity.

Figure 5-7 shows a sample result obtained from this simulation, which shows network latency performance for a video feed from a single platform to a gateway platform as a function of intra-platform IEEE 802.11 transmit power.

This simulation result would suggest that network performance is not significantly improved by increasing transmission power beyond 10 microwatts (IEEE 802.11 nodes are in close proximity to one another). This example makes clearer the type of detailed design decisions that can be aided through network-layer simulation. Another example of application-layer analysis through OPNET network simulation can be found in [126], where the performance of potential VoIP deployments are investigated through OPNET modeling.

TABLE 5-8. Major Features of the WLAN MAC Model

Attribute Name and Description	Values
Number of platforms	5 (25 by 25 meter square area)
	10 (35 by 35 meter square area)
	25 (200 by 200 meter square area)
Data rate (Mbps)*	
Data rate between the sending and receive nodes	1 / 2 / 5.5 / 11
*AP-to-AP node communications, the data rate is 54 Mbps. For control packets, the data rate is 1 Mbps	
Physical layer	
AP-to-AP node communications	Offset Frequency Division Multiplex at nominal transmission frequency of 5 GHZ and receive power threshold of −95 dBm.
	For determining backoff intervals:
	Slot time = 9e-6 sec.
	Minimum contention window = 15 slots
	Maximum contention window = 1023 slots
Otherwise	Direct Sequence at nominal transmission frequency of 2.4 GHZ and following receive power threshold for the corresponding data rate values:
	1 Mbps: −94 dBm
	2 Mbps: −91 dBm
	5.5 Mbps: −89 dBm
	11 Mbps: −85 dBm
	For determining backoff intervals:
	Slot time = 20e-6 sec.
	Minimum contention window = 31 slots
	Maximum contention window = 1023 slots
Transmission power	0.1 watts to 1e-7 watts
Environmental profile	No Environment
	Forest
	Jamming
	Foxhole
	Running
WLAN MAC attributes	
RTS/CTS exchange for medium reservation	
Maximum number of retransmissions	Disabled / Enabled
Fragmentation	7
Roaming/PCF	No WLAN MAC fragmentation
Maximum receive lifetime	Disabled
Buffer size	Infinite
	Infinite

TABLE 5-9. Model Attributes and Their Values

Features	Description
Access mechanism	Carrier sense multiple access and collision avoidance (CSMA / CA) distributed coordinating function (DCF) access scheme as defined in the standard
Frame exchange sequence	Data and acknowledgment frame exchange to ensure the reliability of data transfer. Optional return to send (RTS) / clear to send (CTS) frame exchange for media reservation
Deference and backoff	Interframe spacing DIFS, SIFS, EIFS for DCF. The values of the intervals are selected based on the physical characteristics. Binary exponential backoff.
Data rates	Data rates supported by the WLAN protocol are: 1 Mbps, 2 Mbps, 5.5 Mbps, and 11 Mbps.
Recovery mechanisms	Retransmission mechanism for data frames used when the acknowledgment frame is not received.

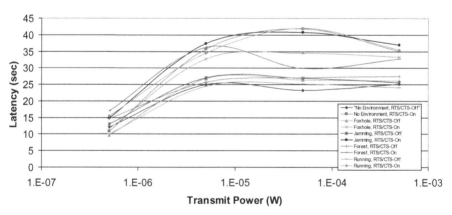

Average Cumulative Latency v. Transmit Power (1 Mbps, Infrastructural Mode, 5 soldier case, low density video)

FIGURE 5-7. Network latency performance as a function of IEEE 802.11 transmit power.

CHAPTER 6

Hardware-in-the-Loop Simulations

Hardware-in-the-loop (HITL) simulations incorporate physical hardware systems as integral parts of the larger simulation system. Employing a HITL approach may be very powerful, depending on the simulation purpose and the role HITL plays in the simulation. For example, a simulation may be designed to test a variety of vendor hardware platforms under the exact same conditions to yield insight into differences in performance for vendor implementations. A properly designed HITL simulation could provide the consistent environment to maintain such conditions so that each platform's performance could be evaluated and compared fairly with its peers. This chapter covers the subject of HITL in detail, including an overview of HITL, engineering tradeoffs, and examples.

As a designer or developer, it may be desirable to develop a simulation system to test system behavior or performance. Typically, simulations consist of a set of inputs, some algorithms and routines to model the system behavior, and a set of outputs. When HITL is employed as part of a larger simulation system, the simulation may consist of more than just a host computing platform to execute simulation code. Hardware components that form part of the simulation architecture may serve a variety of functions within the larger simulation system. Figure 6-1 illustrates a layered representation of a communication simulation, showing explicit simulation processes for the physical, link, network, and application layers.

Any network simulation may encompass more or less layers depending upon the desired functionality to be modeled. In the general representation shown, each layer may pass user data to and from its adjacent layers via the "data plane" (e.g., link layer can pass data to the network layer above or the physical layer below). A "control plane" has been indicated for the purposes of overall simulation control and coordination among each of the layers. This control may be achieved in a variety of ways, including by developing

An Introduction to Network Modeling and Simulation for the Practicing Engineer, First Edition.
Jack Burbank, William Kasch, Jon Ward.

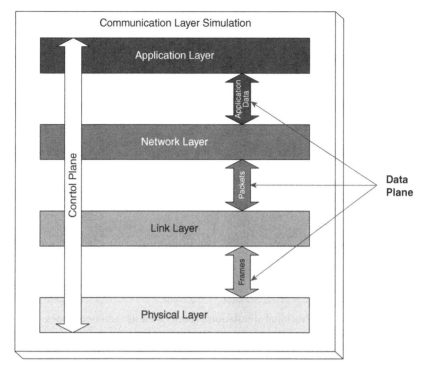

FIGURE 6-1. Layered representation of communication simulation.

standardized interfaces between layers and message format definitions that govern control information exchanges to and from each layer of the simulation. More on this topic is covered in Chapter 7.

In any general network simulation, each instantiated "node" within the simulation may inherit specified functions of any number of layers within the communications stack. Figure 6-2 illustrates a notional configuration with appropriate interfaces between layers for two instantiated "nodes" within a simulation.

Here, two simulated nodes are illustrated, each of which has four layers represented in the simulation: physical, link, network, and application. Simulated data streams flow between each of these layers as would be expected in a layered protocol approach, while simulation processes external to the node instantiation provide some level of control. The simulation processes can control a number of factors, including, but not limited to:

- Application layer events that trigger generation of application layer data
- Network layer address assignments
- Link layer medium access methods (e.g., CSMA, TDMA)
- Physical layer channel conditions, modulation, or coding methods

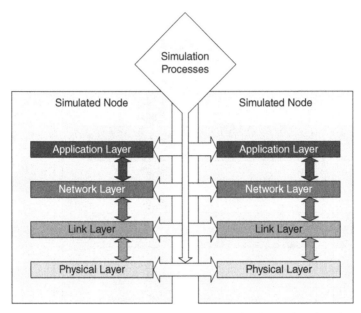

FIGURE 6-2. Notional simulation process with instantiated nodes.

In general, HITL approaches can be employed at any or all of the layers of a particular network simulation. It is important for the designer to consider answering the following questions when considering HITL as part of a simulation solution:

1. What role will the HITL play in the simulation?
2. What are the benefits and costs of employing HITL in this role versus developing a software alternative?
3. What interfaces are necessary between the HITL elements and the simulation engine?
4. What requirements to maintain synchronization will be necessary between the simulation and the HITL device?
5. What inputs are required to the HITL elements?
6. What outputs are desired from the HITL elements?

Table 6-1 summarizes the design choice elements for employing HITL, once a HITL approach has been determined to be the best choice for the designer or developer.

The rest of this chapter will cover advantages and disadvantages of HITL approaches, discuss HITL-specific ideas that could be employed at various layers of the protocol stack, and provide example HITL approaches from the literature.

TABLE 6-1. Design Choice Elements for Employing HITL

Design Choice Element	Description
Role of HITL in the simulation	When identifying the role for HITL, it is important to consider the overall simulation goal—what question is the simulation trying to answer and why is a HITL approach the most appropriate choice for the particular role. HITL may play a key role or an accessory role within a simulation—for instance, a HITL hardware server may be used to assign addresses as a backup to the software-based server simulator. Or, as a more substantial role, a HITL wireless networking radio manufactured by a particular vendor may be tested along with other vendors' radios to determine relative performance. Scalability and cost should be considered in depth here, as the role of the HITL device(s) can drive the cost up substantially if the wrong architecture model is employed.
Interfaces between HITL and simulation	Generally, a control interface will be defined between a HITL device and the simulation engine that will enable the simulation engine to establish control, synchronization, and status as appropriate. However, if simulated data streams are traversing the HITL device, it will be necessary to define data interfaces as well.
Timing	The HITL device(s) will need to operate with particular attention to event update timesteps and other time-driven simulation requirements. HITL devices should always be able to maintain appropriate timing and synchronization in concert with the rest of the simulation. Appropriate interfaces and message exchanges in the control and/or data planes will be necessary to maintain the appropriate level of time synchronization.
Inputs and outputs from the HITL elements	The HITL elements may simply operate in a passive mode where no outputs are required, or may need to provide functions that require regular inputs and output reporting. Inputs and outputs could pass over the data or the control plane, or both.

6.1 ADVANTAGES AND DISADVANTAGES OF HITL APPROACHES

HITL is generally considered to be very powerful because it allows for testing of actual physical hardware (e.g., a wireless radio, or a router) while employing the benefits of a simulation framework that can otherwise maintain a fixed virtual environment to test the hardware; however, this is not the only role suited for HITL approaches. HITL can also be used to model certain functions of the protocol stack that may be otherwise impossible to model with sufficient fidelity in a simulation while maintaining cost and runtime goals for the project as a whole.

HITL for network M&S may be employed at any singular layer of the protocol stack, or may be employed at a combination of layers. The choice of which approach to use is largely determined by the designer of the simulation and the desired speed and cost of the simulation. In general, there is a tradeoff between integrating hardware-only simulation, HITL/hybrid simulation, and software-only simulation. Figure 6-3 illustrates this general trade. Here, the amount of runtime resource is indicated by the y-axis, while the cost to implement is indicated on the x-axis. For software-only solutions, it is generally well known that runtime can be traded for the cost of implementation. For instance, a designer may write a simulation in C to model a network but finds that the simulation does not handle increasing numbers of nodes and heavier traffic loads without a substantial increase in the amount of time the simulation takes to complete. Alternatively, the designer may choose to program a field-programmable gate away (FPGA) to execute the simulation and may find that runtime is greatly reduced for the same set of simulation parameters. Such a method would generally fall close to the "hardware only" simulation box

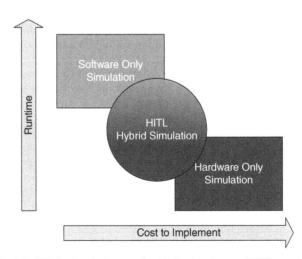

FIGURE 6-3. HITL simulation tradeoffs for hardware, HITL, and software.

indicated in Figure 6-3. Here, the cost to implement would generally be greater than the "software only" approach because of the required additional processes to create a FPGA simulation (the FPGA integrated circuit chip and corresponding circuit board, for instance). HITL approaches generally fall somewhere in the middle of the two extremes.

In order to determine whether a HITL approach is suitable for a particular simulation, the designer must determine the runtime and/or cost benefit to the simulation design effort. Furthermore, it is important to determine where in the simulated protocol stack the HITL will exist, and what kind of a role any HITL will play in the overall simulation. These choices for employing HITL may impact scalability of the simulation, especially where HITL may play a role in modeling a substantial portion of a singular layer (examples will be discussed later). Some HITL may be employed to function as surrogate nodes or subnetworks within a larger internetwork simulation, while other HITL may be employed to function at specific layers, such as a high-speed external DSP that may operate at the physical layer by imparting fading channel models on input time-series complex baseband signals fed by the simulation engine. More detail is covered in the next section on some of the possible HITL approaches.

6.1.1   Advantages of HITL Approaches

Generally, HITL approaches are advantageous in the following ways:

- They allow for testing of physical hardware performance characteristics by precisely controlling the simulation environment—such examples could include conformance and interoperability testing to ensure different vendor compliance with common standards.
- They allow for the possibility of reverse-engineering hardware to determine reactions to external, highly controlled stimuli.
- They allow for decreased execution time relative to a software-only approach (though this is highly dependent on the role of the HITL device(s) in the simulation, and respective capabilities).

6.1.2   Disadvantages of HITL Approaches

Generally, HITL approaches are disadvantageous in the following ways:

- They are generally not well suited for large-scale simulation with many network entities. Depending on the role of HITL in the simulation, scalability may be a major issue. For instance, if a single HITL device is required to represent a particular simulated node, then the number of HITL devices increases in direct proportion to the number of nodes simulated.

- HITL devices may not be readily reconfigurable to consider a variety of different scenarios. One great benefit of simulation in general is the ability to consider a number of different scenarios and play "what-if" games—i.e., change a protocol behavior, modify network topology, or alter network traffic loading. Depending on the role of the HITL device, it may constrain the simulation conditions because of its own configuration and/or performance limitations.
- HITL devices will require appropriate interfaces to connect into the simulation environment to exchange data and/or control messages with the simulation entities, including the simulation engine and the simulated nodes. These interfaces will need to be defined in a clear manner and tested extensively to ensure proper operation of the HITL device(s) in its (their) role in the overall simulation.

6.2 NETWORK M&S HITL APPROACHES

It is important for the designer to consider what role in the simulation HITL will play, as well as which of the layers the HITL will impact. This section discusses some ideas to incorporate HITL approaches into a larger simulation for the three lowest layers related to network M&S: physical, link, and network.

6.2.1 HITL at the Physical Layer

At the physical layer, the following basic functions are performed:

- Transformation of bits of information into a signal representation
- Transmission of the signal
- Propagation of the signal through a channel
- Reception of the signal

HITL can be employed for any (or all) of these functions at the physical layer. For example, a HITL approach could incorporate a channel model that relies on an external DSP to model characteristics of the channel such as small-scale fading. Implementing a small-scale fading channel in software may be much slower than a similar hardware implementation because of the runtime required to model tapped-line filters that are commonly used for this approach. Figure 6-4 illustrates the example of using an external DSP hardware system to perform this function. Here, the simulation engine controls the entire simulation and provides the necessary data translation between the software-only portion of the simulation and the HITL of the simulation. The control plane is indicated to touch both software-only and the HITL portions to maintain synchronization between the two separate

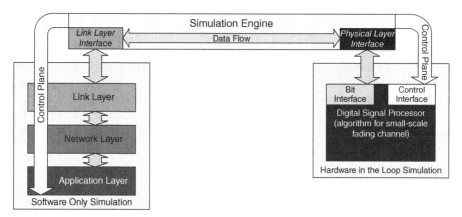

FIGURE 6-4. HITL approach for physical layer channel model with digital signal processor.

simulation entities as well as provide configuration parameters to each layer and associated node instantiations (not shown). In the software-only portion of the simulation, the link, network, and application layers are modeled. Network user data flows in between these layers according to the interfaces between each as they are modeled in the simulation. At the link layer, data are passed directly to the simulation engine via a link-layer-to-simulation-engine interface on the data plane. The simulation engine then performs any translation necessary between the link layer data and the physical layer data, and passes the physical layer data to its physical-layer-to-simulation-engine interface on the right. The physical layer interface here is the interface that passes data between the DSP (via the DSP bit interface) and the simulation engine. The control plane also connects the DSP to the simulation engine so the simulation engine can coordinate and command the DSP to perform specific functions such as clearing its buffer, resetting, and configuration to process the bit stream in fixed block sizes, for example.

Within the DSP, an algorithm exists to process small-scale fading changes to the bit stream on the input portion of the bit interface. While not explicitly shown, the bit interface does include an output that sends processed bits (potentially altered by the fading channel model depending upon modeled conditions) back to the simulation engine via the physical layer interface. These bits are then sent to the link layer interface of the simulation engine for passing to the link layer within the software-only portion of the interface. Notionally, for a given source-destination pair, the source node instantiation link layer bits would be passed to the simulation engine, then passed to the DSP for processing, then passed from the DSP back to the simulation engine. From there, the simulation engine would pass the bits back to the link layer of the destination node instantiation for processing.

The simulation engine would effectively coordinate the bit flow between the virtual nodes.

In this example, it is clear that the DSP HITL performs the role of a small-scale fading channel model for the larger simulation. The rest of the network is modeled in software. Such an example may be quite useful for simulating mobile wireless networks where nodes are likely to experience a high degree of small-scale fading effects. Once again, it is important for the designer to consider what aspects of the network are most critical to model and choose the model appropriately.

6.2.2 HITL at the Link Layer

At the link layer, the following basic functions are performed:

- Medium access control (MAC)
- Error correction and/or detection
- Automatic repeat request (ARQ) or acknowledgment mechanisms

Implementations of particular link layers (e.g., the IEEE 802.11 MAC) can be tested exclusively through HITL methods. Oftentimes, such link layers are developed in software then ported to FPGA or other hardware as part of a larger radio system (e.g., IEEE 802.11-based access point (AP)). Consider the case where a designer may want to test a variety of link layers for a given network simulation to determine which one functions the best for a fixed set of simulation parameters but does not have access to the proprietary code used to instantiate the link layers on the FPGAs. Such simulation parameters could include network topology, mobility profiles, or traffic loading characteristics. Figure 6-5 illustrates this idea for a simulation with two nodes.

In Figure 6-5, only the data plane connections are illustrated. There are two simulated nodes in the network, and where the link layer would normally be simulated is an indicator that connects the simulation engine to the network layer and the physical layer. The simulation engine maintains standardized interfaces to both of these layers as well as an FPGA interface it uses to communicate with the FPGA instantiations of the various link layers under test. It should be noted that each simulated node would likely require its own link layer FPGA—so testing two simulated nodes for Link Layer #1 would necessarily require two FPGAs, each implementing Link Layer #1. In this sense, the designer should pay attention to how easily HITL scales with increasing the number of nodes, especially when modeling singular layers in this way. However, this approach is quite powerful in that it allows the user to objectively compare the three link layers under test within an otherwise equal simulation environment.

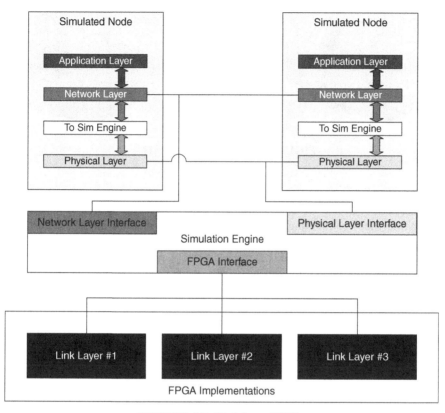

FIGURE 6-5. Link layer HITL.

6.2.3 HITL at the Network Layer

It is interesting to consider cases where HITL is employed at the network layer. This is one case where the role of the HITL becomes increasingly important. The basic functions of the network layer include:

- Logical addressing
- Routing
- Fragmentation/reassembly
- Error handling/diagnostic information

HITL approaches could focus on any one of these functions, and as such, the role within the simulation could vary substantially. For instance, a HITL approach could implement a DHCP server on an external host connected to the simulation engine through a network interface to handle assignment of IP addresses to simulated nodes, or an external router could be placed as a HITL

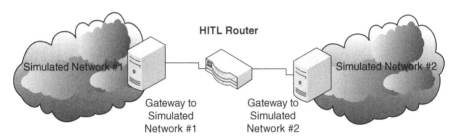

FIGURE 6-6. Network layer HITL.

element between two separate simulated networks with standardized network layer interfaces connecting the router to both simulated networks and their associated "gateway" simulated nodes. Figure 6-6 illustrates the idea for employing a router as a HITL element in this way.

In Figure 6-6, two simulated networks are shown, indicated by the clouds to the left and right. Within each simulated network, there is a gateway "node" identified that provides a standardized interface between the simulated network and the HITL router. The HITL router is located between the two simulated networks and in this role, would serve as an intermediate router between the two separate simulated networks. In reality, if each simulated network is running on a single host computer, it is likely that the interface to the router would simply be another network interface card (NIC) in that host. The simulation would then be able to access the NIC as another "virtual" node within the simulation, identified as a gateway to the router external to the simulation. To test full router functionality, a designer could then instruct the router to be the default router between Simulated Network #1 and Simulated Network #2, and create traffic flows between virtual nodes in each simulated network. So long as the simulated networks implemented routing protocols at the gateway nodes that were compatible with the HITL router protocols supported, packets could be exchanged at the network layer between the two networks and performance of the router could be assessed for a variety of simulated scenarios. This kind of configuration may be useful in testing how different router vendor implementations perform when operating with various versions of routing protocols implemented in simulation. The HITL approach is powerful here because it allows actual testing of real hardware—all while maintaining the ability to keep the virtual simulated environment stable and constant for all the HITL routers under test.

6.2.4 Generalized Approach to HITL at Any Layer

Previously some ideas for employing HITL at the physical, link, and network layers were discussed; however, a generalized simulation framework can be defined for employing HITL at any layer. This section focuses on presenting

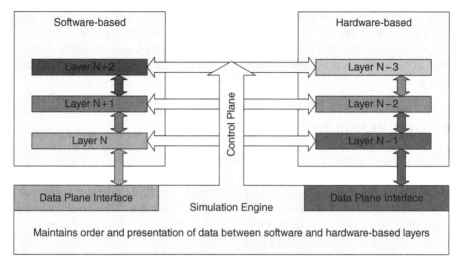

FIGURE 6-7. General HITL abstraction at any layer.

this generalized framework as a starting point for a designer who may be interested in HITL approaches in the simulation.

Generally, for a simulation with some arbitrary number of layers to be modeled, there are some common best practices that can be employed. As has been explained previously, there is a control plane for the simulation and a data plane for the simulation. The control plane generally carries messages between the simulation engine and the various entities in the simulation (instantiated nodes and their associated layer representations). The data plane is where the actual application data are transferred to and from simulated nodes, and passed between layers within a simulated node. The simulation engine maintains synchronization between all the entities and may act as an event trigger, among other roles. Figure 6-7 illustrates this general framework.

In Figure 6-7, there are three distinct layers in the simulation that are software-based and three distinct layers that are hardware-based. A control plane from the simulation engine connects to all layers to maintain simulation control and management. The simulation engine also maintains the order and presentation of data bits between the software-based and hardware-based layers. In the figure, Layer N presents data to Layer N-1 by passing the data to the simulation engine through a standard data plane interface that is common to both Layer N and the simulation engine. The simulation engine then maintains the order of the data and presents the data via another data plane interface standard to both the simulation engine and the hardware-based Layer N-1. While the flow illustrated indicates only one boundary where the data plane traverses the simulation engine, there could be a mix of software-based and hardware-based layer implementations that may require more than one boundary crossing between software- and hardware-based layers. In this case, it is very important for the simulation engine to maintain the correct direction and flow of data across the layers, and as such will increase the complexity of the simulation engine to maintain this flow.

6.3   HITL EXAMPLES

This section provides an overview of some specific examples in the literature that have employed HITL techniques. The purpose of this section is to provide the designer/developer with real-world cases where HITL has been employed, while discussing the benefits and limitations for each case. Two examples are presented here: a wireless networking testbed employing HITL, and a military networking testbed employing HITL.

6.3.1   HITL in a Wireless Networking Testbed

The Johns Hopkins University Applied Physics Laboratory (JHU/APL) developed a HITL testbed to simulate fast-moving mobile nodes [69]. This testbed was affectionately named "ACTION," which stands for "Adaptable Channel Testbed for Investigation of On-the-move wireless Nodes." By employing a variety of software and hardware platforms, ACTION can model a variety of environments to test wireless networking radio performance. Here, the HITL is a wireless radio such as an IEEE 802.11 AP or client card, along with some RF components such as signal attenuators, phase shifters, and frequency translators. Figure 6-8 provides an overview of the ACTION simulation.

In Figure 6-8, the purpose of ACTION is to provide a simulated wireless network environment (SWNE) by utilizing amplitude, phase, and frequency control (APFC) modules driven by LabVIEW hosted on a PC platform. These APFC modules can change the signal characteristics between any base station and the mobile node. The signal between the base station and mobile node is represented mathematically by the following equation:

$$S(t) = A(t)\cos(2\pi f(t)t + \phi(t))$$

Where $S(t)$ is the total composite signal, $A(t)$ is the amplitude of the signal as a function of time, $f(t)$ is the frequency of the signal as a function of time, and $\varphi(t)$ is the phase of the signal as a function of time. Each of these parameters can carry information and is likely to change in a mobile environment. This representation of the signal drove the choice of using APFC modules.

For a specific scenario definition, say, for a mobile node moving through a particular wireless coverage area, each of the "base stations" (or access points) may be set up as a network access point-of-presence within the SWNE. The SWNE is driven specifically by the APFC modules that exist between each of

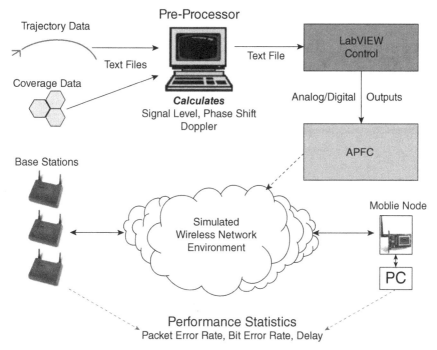

FIGURE 6-8. HITL Wireless Networking Testbed: JHUAPL ACTION [69].

the base stations and the "mobile node" or client card. Trajectory and coverage data are fed into a pre-processing platform as part of a scenario definition. These data drive signal level, phase shift, and Doppler or frequency shifting characteristics for a given mobile platform as it moves through the SWNE "coverage area." In reality, signal levels, phase, and frequency are being changed incrementally between each of the base stations and the mobile node across a simulation run for a defined scenario. It is expected that the mobile node will generally associate with whatever base station has the most robust signal level. Performance metrics can be collected throughout the simulation execution to record how performance changes as the scenario evolves. Such metrics for a wireless network system could include the PER, BER, goodput, throughput, and latency as described in Table 1-2.

Figure 6-9 illustrates the input and output sets for the ACTION simulation pre-processor. Here, the number of base stations and their coordinates define the coverage area for the scenario. Antenna patterns are provided as inputs for the base stations to determine how the signal strength would vary as a function of latitude, longitude, and altitude. Such information is useful in providing a simulation environment that models changes in the mobile node position relative to the base station. A mobile trajectory is defined for the mobile node, in a (latitude, longitude, altitude, time) format. The wireless

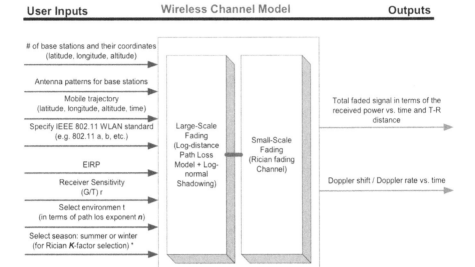

FIGURE 6-9. ACTION inputs and outputs for the pre-processor [69].

standard used is provided (in this example, IEEE 802.11). The effective iso-tropic radiated power (EIRP) is defined for the transmitting radios (both the base stations and the mobile node), and the receiver sensitivity is defined for each as well. The environment is selected as an input (which drives the propa-gation model choice in terms of the path loss exponent for the large-scale fading model), and the season (e.g., summer or winter) is chosen (for the Rician K-factor that drives the small-scale fading model). Two outputs are given: one is the total faded signal level in terms of the received power versus time and transmitter-receiver distance, and the other is the Doppler shift and rate versus time. These outputs then drive the APFC modules that, for a given simulation run, will precisely control the amplitude, phase, and frequency of each of the base station-mobile node links over the course of time for the simulation run.

Figure 6-10 illustrates the network topology algorithm for ACTION. Here, when the number of network nodes (base stations) is defined, the placement of these stations is chosen at random or in specific locations to form the basis of the simulated wireless environment. If specific locations are chosen, they are specified on the basis of their latitude, longitude, and altitude. Automated decisions for random placement can take into account the effective radio range to cluster base stations close together or further apart depending on the scenario type desired.

Figure 6-11 illustrates the mobile node trajectory definition process for ACTION. Here, a mobile node's trajectory may be specified as a random or specific trajectory; if the trajectory is specific, it needs to be defined by its

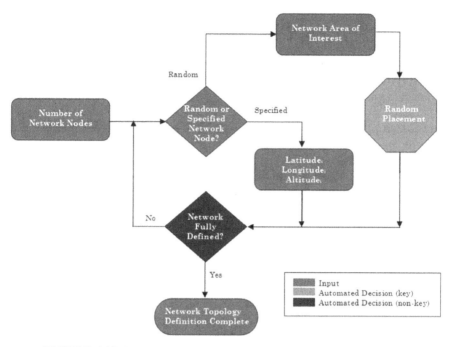

FIGURE 6-10. Network topology definition process for ACTION [69].

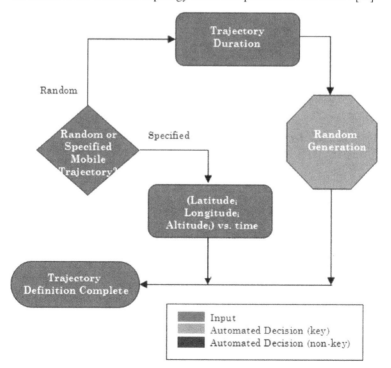

FIGURE 6-11. ACTION mobile node trajectory definition process [69].

(latitude, longitude, altitude) versus time. If it is a random trajectory, then the duration of the trajectory must be specified and a random generation of a path takes place.

Figure 6-12 illustrates the actual physical hardware that is implemented for the ACTION simulation. Here, each of the base stations on the left are interconnected via a network switch so they connect to the same network resource (source or sink) to enable the mobile node to access the same resource (if end-to-end performance is being measured). Each of the base stations has a RF antenna port that is hard-wired directly into the APFC module. Within an APFC module, there are two signal attenuators providing a dynamic range of around 80 dB of signal attenuation strength to model small- and large-scale fading effects, a frequency translator that uses a serrodyne method to shift the frequency to model the Doppler shift, two circulators that split transmit and receive paths to enable separate phase shifts from each path to model small-scale random phase shifting (largely expected through multipath interactions), and two directional couplers for transmit and receive paths, respectively, to measure the affected signals for verification, validation, and diagnostic purposes. On the right, the mobile node is shown. Here, the mobile node

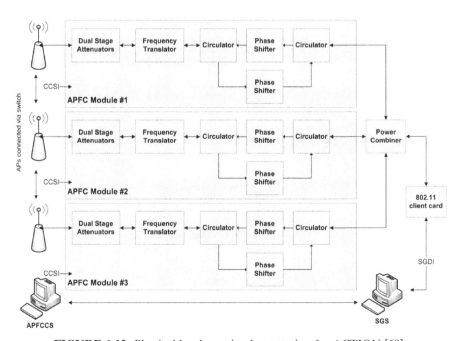

FIGURE 6-12. Physical hardware implementation for ACTION [69].

receives the signals from each of the base stations via a power combiner circuit, while simultaneously sharing its transmitted signal with each of the three base stations. The frequency shift and attenuation are equal in both directions as the link characteristics are assumed to be constant. The "CCSI" indicated stands for computer control signal interface and is the interface between the APFC devices and LabVIEW. Analog voltages connected to pin inputs on each device control the APFC devices. LabVIEW drives each of the devices by controlling the analog voltages via an analog output card.

Note that for every base station, there is a required APFC configuration that must be instantiated to provide a precise controlled environment to model the link between that base station and the mobile node. The number of APFCs required in a simulation is the number of mobile nodes times the number of base stations:

$$N_{APFCs} = N_{MNs} * N_{BSs}$$

As such, it is necessary to purchase all the components of each APFC for an arbitrary configuration where there are N_{MNs} and N_{BSs}. Each of the components here can be quite expensive depending on the frequency range, dynamic range, and other parameters that drive cost of design for the individual components. In this sense, a HITL simulation based on the ACTION implementation may not scale to many nodes and another solution may need to be considered if the desired simulation is to model many nodes in a cost-effective way.

Figure 6-13 illustrates an example product from the ACTION simulation. Here, the effects of frequency shifting are shown for the 1 Mbps IEEE 802.11b waveform. Specifically, it is shown that the PDR illustrated on the y-axis increases as the magnitude of the frequency shift increases. Eventually, once the frequency shift magnitude exceeds about 40 kHz, the mobile node client card loses association with the base station. This is one example of ACTION being employed to investigate the issue of Doppler shifting on the IEEE 802.11b waveform.

An overview of the ACTION simulation has been presented here. ACTION seems to provide an advantage over modeling the IEEE 802.11 waveform in software in that it allows for testing hardware directly (base stations/access points and wireless client cards) with in-line RF components that modify amplitude, phase, and frequency to create a wide variety of simulation scenarios. By testing the hardware directly, it is expected that the simulated environment will yield insight into performance once the hardware is employed in a similar real-world scenario. However, it is clear that once the number of nodes modeled increases, a corresponding increase in the number of APFC modules is required to support the simulation environment modeled in the ACTION simulation architecture. Because each of the modules may be

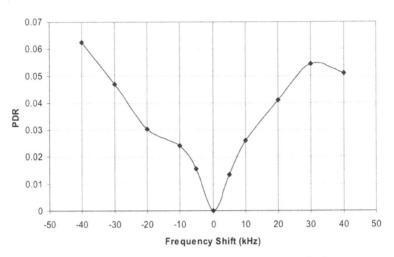

FIGURE 6-13. Example result from ACTION [69].

expensive to procure, the particular implementation presented may not be well suited for large-scale simulation of many wireless nodes.

6.3.2 Army HITL Using OPNET

An overview is provided in [70] of a U.S. Army-sponsored initiative to use HITL techniques with the OPNET modeling suite to support developmental testing of the Joint Tactical Radio System (JTRS). JTRS is a program of record for the U.S. Department of Defense (DoD) that is focused on developing radio platforms for soldiers to use in the field. The radios will generally support some network functions. The particular objective of the HITL-OPNET model described in [70] focused on developing a capability to evaluate the performance of large, complex Army communications networks by combining both virtual models and physical implementations of real networks. By employing this approach, the Army modelers can play "what-if" games with network configurations in the simulation and see how they impact the physical network hardware connected, or vice versa. Figure 6-14 illustrates one example for utilizing the OPNET modeler system to test real JTRS network hardware as part of a "network-in-the-loop."

In Figure 6-14, JTRS network models have been defined in the OPNET modeler and are connected to the Internet (or through a virtual private network tunnel, depending on the security requirements) via exit routers. Each of the exit routers is defined in the OPNET modeler to be a physical Ethernet

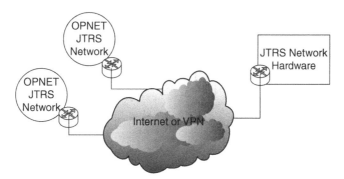

FIGURE 6-14. JTRS developmental test [70].

interface on the host platform that contains the OPNET model. The OPNET system-in-the-loop (SITL) module is employed to provide the necessary translation between the simulation environment and the physical Ethernet connection to the Internet on the host platform. The JTRS network hardware is connected to the simulation through the network connection and can literally send and receive packets to and from the OPNET simulation and participate as though it were literally part of the simulated network. This idea is similar to the idea presented in Figure 6-6 in that the connection between the physical hardware and the simulation is at the network layer. The OPNET model can be used to collect statistics on the overall network. The JTRS network hardware can be stimulated via discrete event processes within the OPNET model or can generate its own events triggered elsewhere.

The Internet/Virtual Private Network (VPN) connectivity allows the JTRS network hardware to be connected anywhere in the world where an Internet connection exists regardless of where the other simulated networks are hosted, so long as they too are connected to the Internet; however, such an approach should always consider a security analysis to determine the vulnerabilities that may exist when connecting such hardware to the Internet. A VPN approach may mitigate this problem, but any designer or developer should always be concerned when connecting up any equipment (simulation or hardware) to the Internet, especially if sensitive information could be exposed. An appropriate security protocol or suite of protocols may need to be employed to provide sufficient security levels to prevent unauthorized access to any part of the simulated system.

6.3.3 HITL Network Simulator for Large-Scale Military Wireless Communication Systems

In [71], the authors presented a HITL network simulator to analyze specially developed military network hardware. In particular, the authors developed hardware based on the Multi-Role Tactical Common Data Link (MR-TCDL),

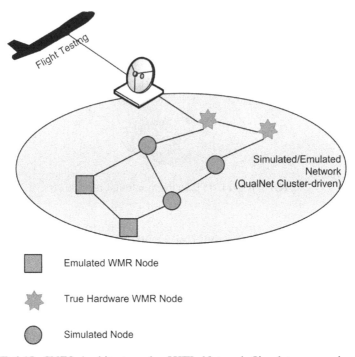

FIGURE 6-15. CNES Architecture for HITL Network Simulator reproduced from [71].

which is a wireless MANET system, in conjunction with their Inter-platform Communications Manager (IPCM) software. Together, these two technologies form the basis of the platforms of interest for testing in the larger HITL simulation. The authors were particularly interested in modeling using the HITL approach because they needed to test their hardware operating in a large-scale network environment. Furthermore, they were required to flight test their hardware solutions and needed access to a large-scale simulation as flight testing would be prohibitively expensive. The authors developed a HITL architecture known as the Communication Network Effects Simulator (CNES), illustrated in Figure 6-15. Here, WMR stands for "Wideband Mobile Router" and is the hardware platform the authors developed to test functionality. The WMRs in their simulation architecture were a combination of physical hardware devices and emulated devices. The system was implemented using the QualNet network simulation tool, with the IP Network Emulation (IPNE) module. The IPNE module provides similar functionality in QualNet as compared to the SITL module for the OPNET modeler. The IPNE module allowed the simulation environment to interface directly to hardware WMRs and emulated WMRs. The authors claim that the scale of the network could be greatly increased using this method to exercise functionality of their WMR platform.

Using their simulation platform in pre-testing, the authors found several issues with their WMR platforms that they were able to fix, related to closed loop power control, directionality of data links, and Simple Network Management Protocol (SNMP) management of devices. Furthermore, they were able to diagnose memory problems in the hardware implementations of their IPCM code on the WMR platforms by being able to detect unstable behavior during simulation execution. During real-time validation conducted while the simulation was executing, the authors found that the CNES system helped determine maximum propagation range targets and helped resolve issues found during the node discovery process of the MR-TCDL system. Furthermore, their testing found several issues with the hardware implementations that would have impeded successful flight testing.

The CNES system was designed primarily to test developed hardware platforms in the presence of a large-scale networking environment to help mitigate risk related to further development, fielding, and flight testing. The authors in [71] developed a network simulation system that interfaced at the network layer between the simulated network in the QualNet model and the physical hardware and emulated WMR platforms. The advantages to mitigating risk for this program seem obvious—flight testing hardware is quite expensive and their hardware was required to operate in a large-scale network environment; however, it should be noted that the degree of fidelity achieved by using the QualNet modeler should be taken into account when designing any simulation to achieve large scale. Large-scale networks can be approximated in simulation by aggregation of network-layer traffic (especially if there are key gateways and highly asymmetric traffic patterns). The authors used a distributed simulation approach to increase their computing power to model the large-scale network in an appropriate way. It is desirable for a network designer or developer to take these factors into account when designing a simulation that focuses on providing a large-scale behavior such as the one in this example. Note here that the role of the HITL devices (the WMRs) is not to model the large-scale network—rather, that role is relegated to the QualNet software model. Here, that model simply helps replicate the expected environment that the WMRs will likely operate in—one where there are a substantial number of network nodes and a large-scale network topology exists. The HITL roles are focused specifically on functionality related to the WMRs—supporting testing of their functions in a large-scale environment to minimize the risk associated with flight testing and to further their development.

6.3.4   HITL with IEEE 802.16e Devices

IEEE 802.16e is a wireless networking standard that enables cellular-like communications between subscriber stations (SSs) and base stations (BSs). Unlike its predecessor IEEE 802.16d, mobility is supported, so BSs can coordinate SS handovers. In IEEE 802.16e, the BS coordinates all transmissions. It may be of interest to evaluate handover performance at the PHY and MAC layers for IEEE 802.16e (also referred to as mobile WiMAX), and this section covers an example where physical hardware could be used as part of a larger simulation to evaluate this type of behavior.

Figure 6-16 illustrates an example for HITL applied to mobile WiMAX hardware. Here, two base stations form a virtual network via variable attenuators used in combination with a power combiner and a single subscriber station. There are multiple independent paths that need to be established to ensure the proper connections are made between the various components that make up the HITL simulation. In Figure 6-16, there are four primary paths shown, each in a different color. The mobile WiMAX base stations (indicated in orange) connect to both a content network (green) and a hard-wired RF

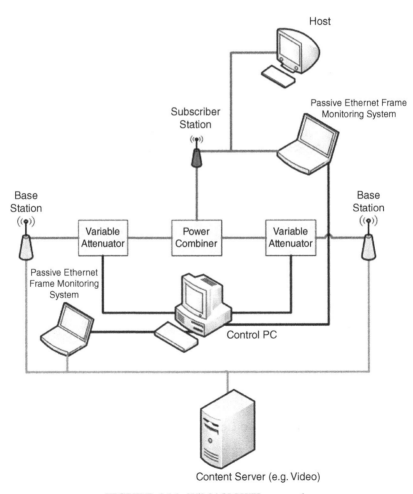

FIGURE 6-16. WiMAX HITL example.

path (orange). On the content network, the base stations can coordinate handover and also connect to a content server (e.g., video server). On the RF path, each base station is connected to a variable attenuator that can be set to a wide range of attenuation values. The output of each attenuator is connected to a power combiner that combines the attenuated signals from both of the base stations to the subscriber station. The effect of the variable attenuator allows the control PC to simulate a wireless networking environment where the subscriber may see a higher signal level from one base station compared with another and then undergo a relative attenuation change and observe the opposite. The control network (indicated by red lines) provides the necessary out-of-band control for the simulation, driven by a control PC. The control PC is connected to both variable attenuators to actively change attenuation over time driven by some predefined script. The control PC is also connected to two passive Ethernet frame monitoring systems, which are generally laptops that run Wireshark to capture traffic for both the content network and the host network (purple). The traffic captured from these laptops can be analyzed to assess performance of handover (by following packet exchanges between the base stations during handover events) and the content server stream. The host is connected to the subscriber station directly and uses the subscriber station to access the content server that is connected to both of the base stations.

In this example, multiple experiments can be conducted to evaluate performance of the mobile WiMAX network. Metrics that could be collected include the latency of handover, quality of video streaming under variable attenuator conditions, and the stability of link persistence when both base station signals are received at the subscriber station with the same signal strength. The control PC can be used to execute profiles that drive the variable attenuators to simulate a wireless network environment (similar to the ACTION example shown in Section 6.3.1).

While it is easy to illustrate the design concept of a mobile WiMAX testbed, actual implementations will require a more detailed approach that considers the actual hardware being used within the testbed. The hardware capabilities of each of the components will have specific interface types and processing limitations that will affect which metrics can be collected (and how often), how the hardware will be interconnected, and what overall architecture should be employed to control the entire HITL system. For this example, Table 6-2 summarizes some of the parameters that need to be considered for the various components in the mobile WiMAX HITL simulation.

Table 6-2 summarizes some of the key questions that should be asked when selecting and configuring hardware for the mobile WiMAX HITL testbed. Generally, once these questions are answered, the designer/developer should have a clear idea of all the supporting equipment (wires, cables, and adaptors) and configurations (software) needed to achieve the goals of the testbed. Furthermore, the architecture of the testbed will emerge from such an analysis.

TABLE 6-2. Summary of Parameters to Be Considered in Mobile WiMAX HITL Simulation

Component	Implementation Aspects to Consider

Mobile WiMAX Base Station	• Data network
	  ◦ Interface (Ethernet? RS232?)
	• RF network
	  ◦ Frequency (2.5 GHz, 3.5 GHz, 4.9 GHz?)
	  ◦ Connector type (N, RP-TNC, BNC?)
	• Configuration
	  ◦ Web, telnet, SSH?
	  ◦ Applied over RF interface or data network interface?

Mobile WiMAX Subscriber Station	• Host network
	  ◦ Interface (Ethernet? RS232?)
	• RF network
	  ◦ Frequency (2.5 GHz, 3.5 GHz, 4.9 GHz?)
	  ◦ Connector type (N, RP-TNC, BNC?)
	• Configuration
	  ◦ Web, telnet, SSH?
	• Applied over RF interface or host network interface?

Host	• Operating system (Mac OS X, Windows, Linux?) • Application requirements to access content server (video CODECs, players, web browsers?) • Hardware requirements (processor speed, RAM, hard drive space?) • Network interface (Ethernet? RS232?) • Remote configuration capability?
Control PC	• Operating system (Mac OS X, Windows, Linux?) • Application requirements to drive simulation (LabVIEW, scripting?) • Hardware requirements (processor speed, RAM, hard drive space?) • Network interface (Ethernet? RS232?) • Other interface types (connection to ariable attenuators?) • Software to drive simulation (using existing applications or developing new ones?)
Variable attenuator	• Rated frequency range of operation (2.5 GHz, 3.5 GHz, 4.9 GHz?) • RF interface connector type (N, RP-TNC, BNC?) • Control interface (RS232, Ethernet, voltage-driven?)
Power combiner	• Rated frequency range of operation (2.5 GHz, 3.5 GHz, 4.9 GHz?) • RF interface connector type (N, RP-TNC, BNC?)

TABLE 6-2. Continued

Component	Implementation Aspects to Consider
Content server	• Operating system (Mac OS X, Windows, Linux?) • Application requirements to serve data (video and/or audio server software?) • Hardware requirements (processor speed, RAM, hard drive space?) • Network interface (Ethernet? RS232?) • Remote configuration capability?
Passive Ethernet Frame Monitoring System	• Operating system (Mac OS X, Windows, Linux?) • Application requirements to monitor Ethernet frames on the network (Wireshark, tshark?) • Hardware requirements (processor speed, RAM, hard drive space?) • Network interfaces (Ethernet? RS232? How many?) • Remote configuration capability?

6.4 COMMON PITFALLS FOR HITL APPROACHES

This section provides some of the common pitfalls when employing HITL approaches in a general simulation. It is important for the designer or developer to keep these pitfalls in mind when considering HITL. The pitfalls associated with HITL are summarized in Table 6-3.

6.5 NETWORK-LAYER HITL-READY NETWORK SIMULATION PLATFORMS

There are a variety of HITL-ready network simulation platforms available today. Specifically, two commercial platform modules are discussed in this section: one for the OPNET discrete-event simulation platform and the other for the QualNet networking simulation platform. Any simulation design that incorporates either of these simulation platforms while desiring HITL functionality at the network layer (e.g., for testing hardware devices that need to interact with other network-layer entities) could potentially use the modules discussed in this section. For more detail on each of the modules, the reader is encouraged to contact the respective companies.

6.5.1 OPNET Modeler and SITL

OPNET Corporation has developed a module known as System-in-the-Loop (SITL), which provides an interface to connect physical network hardware or

TABLE 6-3. Common Pitfalls to HITL Approaches

Issue	Description
Employing HITL in the wrong role with respect to the simulation goals and architecture	By choosing the wrong role for HITL in the context of simulation goals and architecture, unnecessary constraints may be placed on the entire simulation to accommodate the HITL devices. These constraints could include limiting processor availability to service/manage HITL devices, or could include having to re-write software within the software-only side of the simulation to accommodate required inputs/outputs to/from the HITL devices. It is important for the designer/developer to consider the role of HITL rigorously and judiciously.
Insufficient scalability	HITL devices come with their own set of capabilities and limitations. Depending on the role of HITL, scalability may not be achievable if the simulation scale (increased number of nodes or traffic load, for instance) needs to be increased. Upgrades to HITL devices may be required and may be more costly than an original software-only approach.
Limitations for future technology enhancements	HITL devices may or may not be equipped with the ability to upgrade their capabilities. For instance, wireless radios may be able to update firmware through a software load or may require hardware updates to support future enhancements. If the simulation architecture calls for upgrading capabilities in the future and the HITL may play a role in providing such capabilities, it is important to consider what upgrade paths are available to the HITL devices from logistical and cost perspectives.
Lack of technical support and training for HITL elements	Depending upon the vendor that provides the HITL elements and/or the complexity of the HITL chosen, technical support and training may be required for proper integration into the larger simulation, especially during troubleshooting phases. Consideration should be given to this effect, depending upon the complexity of the HITL and the implementer's familiarity with HITL integration.

TABLE 6-3. Continued

Issue	Description
Insufficient automation of HITL into the simulation system	Sometimes HITL devices may not be able to be readily configurable from a remote interface and may need manual configuration for particular simulation runs. If the designers of the simulation desire many executions of the simulation to parametrically analyze behavior, maximum automation of the HITL configuration should be a part of the design to minimize the requirement of the human-in-the-loop.
Inappropriate choice of timestep for HITL integration	The timestep functions that define a simulation's ability to step through time and trigger events needs to be synchronized with the abilities of any HITL device. Optimally, HITL should not impede the simulation tempo that could otherwise be achieved with software-only approaches—on the contrary, it should improve the ability of the simulation to achieve goals (by enabling faster runtime and more simulation results per runtime, or providing more fidelity for particular simulation functions).

software applications to an OPNET simulation [72]. This section provides a brief description of the capabilities of this module for OPNET users.

The SITL module provides an interface that enables a physical network hardware device to be connected to the OPNET simulation. To date, it supports both connections via Ethernet and wireless local area network (IEEE 802.11). Any host platform executing the OPNET software can connect to multiple physical network devices via different physical Ethernet or wireless network interfaces on the host. The OPNET simulation instantiates a gateway entity for each of the physical devices to enable packets to be transferred between the simulation and the physical device by providing the necessary translation between the Ethernet or wireless local area network frames in the physical hardware and the simulation packets in software. During the execution of an OPNET simulation using the SITL module with associated physical network devices connected via the gateway entities, packets are able to flow between the simulation and the physical devices according to the simulation scenario and event-driven processes. Conversion modules may be necessary to provide interoperability between physical devices and simulated devices (e.g., routers), depending on the protocol.

6.5.2 QualNet IP Network Emulation (IPNE)

The QualNet network simulation platform [16] contains a feature called IP Network Emulation (IPNE) that is similar to the OPNET SITL module in that it enables connection between the simulation system and physical hardware devices. It allows IP-layer connectivity between the physical devices and the QualNet simulation, so that the physical devices can participate in the simulation scenario. The IPNE interface can control the simulation clock to maintain proper synchronization with the real-time clock on the physical interfaces. Furthermore, it can identify packets that need to be passed between the physical devices and the simulation devices and conducts processing as needed in the simulation environment.

6.6 HITL CONCLUSION

Any designer or developer who is interested in employing HITL for network modeling and simulation needs to consider what role HITL will play in the simulation and why it is important to employ in the context of achieving the simulation goals. It is important to consider how the HITL will interface to the existing simulation and what control is required to maintain synchronization between the HITL platform(s) and the simulated entities. Designers should take care to ensure that the HITL platforms do not inhibit simulation capabilities compared to an all-software approach. Furthermore, if scalability is desired to support future simulation runs or scenarios, it is essential to assess what required hardware enhancements are necessary for HITL devices to support future capabilities. Above all, the designer needs to consider not only the capabilities that HITL provides the current simulation in achieving the goals related to specific simulation functions, but how the HITL impacts the entire simulation architecture, from interoperability and compatibility needs to scalable requirements driven by expected future uses for the simulation platform.

For network modeling and simulation, there are a variety of tools already available that can assist designers and developers in integrating network hardware with existing popular modeling suites such as OPNET and QualNet. However, designers should take care to consider the cost and maintenance of modules for these suites that enable such functionality compared to in-house development cost. Optimally, any good simulation design will take into consideration both hardware and software approaches and choose the one that best meets the simulation goals while minimizing other factors such as cost, runtime execution speed, and implementation timeframes.

CHAPTER 7

Complete Network Modeling and Simulation

To develop a complete network simulation, designers and developers will likely want to employ models at each of the various communications layers that are germane to their specific problem. As such, how do we put together all the different models and simulation tools at the various layers of the protocol stack to provide a complete end-to-end network model or simulation? By employing a multi-layer approach, designers and developers will find that fidelity and complexity will likely increase, as compared to simulating a single communications layer of a network. This chapter discusses the subject of complete network M&S. In particular, examples are provided and simulation platforms that attempt to provide these capabilities are discussed, along with advantages and disadvantages to each. Furthermore, an in-depth discussion of the IEEE High Level Architecture (HLA) standard 1516 is provided as a reference for the designer or developer. Figure 7-1 illustrates an overview of network simulation. Generally, a complete network simulation will include models that provide functionality at the physical, link, network, transport, and application layers.

Chapter 6 covered how HITL approaches could be employed in network M&S, but similar concepts are useful in determining how to create a complete network simulation. For instance, it is essential when modeling multiple layers to define standardized interfaces, both in the data plane and the control plane, to enable the layers to communicate appropriately and pass data and control information to each other. Furthermore, it is important to develop a control engine that drives the entire simulation and maintains appropriate control interfaces to the appropriate entities. While it is up to the designer to determine which entities require control and to what extent, interfaces are required to enable disparate parts of the simulation to meaningfully interact and achieve the overall goal of the simulation.

An Introduction to Network Modeling and Simulation for the Practicing Engineer, First Edition.
Jack Burbank, William Kasch, Jon Ward.
© 2011 Institute of Electrical and Electronics Engineers. Published 2011 by John Wiley & Sons, Inc.

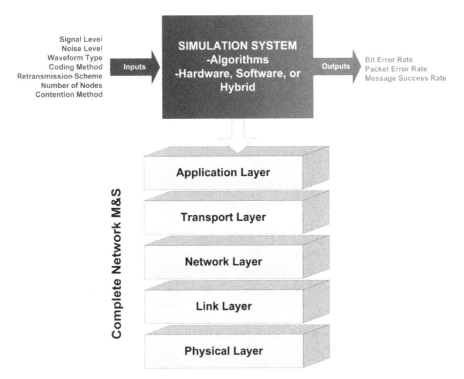

FIGURE 7-1. Overview of network simulation [1].

When the designer considers developing a complete network model/ simulation, it is important to ask the following questions to start:

1. Which layers of the protocol stack will be modeled?
2. How many network nodes will be simulated?
3. What specific protocols and functions must be implemented at each of the layers?
4. How will the control interfaces of the simulation be defined?
   a. Between layers and the simulation engine?
   b. Between nodes and the simulation engine?
5. How will the data interfaces of the simulation be defined between layers?
6. What level of simulation control is appropriate for each of the layers?
7. What level of simulation control is appropriate for each simulated node?
8. What is the timestep configuration of the simulation?
9. Will the simulation be developed to execute on a single platform or in a distributed manner?

10. Will the various functions of the simulation be developed to execute as a set of federates or as a single entity?
11. What functions of the simulation will be developed in-house and which will be obtained off-the-shelf?
12. What plan will be established to validate simulated behavior to ensure results are correct?
13. What results will be desired as outputs from the simulation?

All of these questions can serve as a starting point to give the designer an overview of the types of behavior that must be modeled; however, it is important to note that a variety of engineering tradeoffs may need to be conducted to determine the impact of particular design choices. For instance, a developer may lean towards absolute control of all simulation events (from a node sending a message to a protocol stack reassembling a fragment) via notices sent from the control plane from any simulation entity to the simulation engine. If this is the case, it is very likely that the simulation engine may be overwhelmed with control messages once the limits of simulation scale are tested (through increased traffic loads or an increased number of nodes, for example). Table 7-1 summarizes the elements constituting the design choices for a complete network simulation.

7.1 COMPLETE NETWORK M&S PLATFORMS

There are a variety of network simulation tools available in both the commercial domain and the public domain that provide designers the ability to create a complete network simulation. This section discusses some of the currently available tools. Table 7-2 summarizes some of the tools available, with the website of each tool located below the name. Designers and developers are encouraged to browse these websites and connect with the companies and/or communities that use these tools to determine their feasibility for a particular application.

7.2 IEEE HLA (1516)

The IEEE HLA (1516) standard is a generalized framework for developing and executing distributed simulation systems. By employing HLA, different simulations, referred to as "federates," can intercommunicate, independent of their respective host computing platforms. An overview of an HLA federation is provided in Figure 7-2. Here, different federate simulations are interconnected to a common runtime infrastructure (RTI) within a single HLA federation. The RTI provides the common interfaces to each of the federates and enables information to be exchanged between them while maintaining overall

TABLE 7-1. Design Elements for Complete Network Simulation

Design Choice Element	Description
Number of layers	Generally, if a simulation requires multiple layers, it will also require interfaces to be defined to connect each layer to the simulation engine for control (if necessary) and an interface to pass data up and/or down the stack.
Number of simulated nodes	Larger number of simulated nodes is generally considered better for a complete network simulation, but this is highly dependent on the problem to be solved. Generally, as the number of simulated nodes increases, the overhead required to maintain individual node instantiations increases, increasing runtime.
Functionality at each layer	Depending on the problem to be solved, designers may pick and choose which of the key functions at each layer are modeled. Generally, as more and more of these functions are modeled, runtime suffers but fidelity increases.
Level of simulation control	The level of simulation control is defined as the necessary steps to establish and maintain simulation control. Simulation control includes triggering of events (such as starting and stopping simulation, instantiating nodes, setting initial conditions, and updating conditions). Depending upon the level of control required by the designer, increased control ensures better synchronization, but at the expense of runtime and complexity. Complexity increases in the form of more interface definitions to enable control message exchange between simulation entities and the simulation engine, as well as the algorithms in the simulation engine that process the messages.
Simulation time step	Generally, a simulation will step through time, triggering events and interactions between the entities and recording results in some orderly fashion. The timestep increments will need to be defined here, along with the appropriate interactions that happen at particular timestep increments (such as location updates, message transmissions, and data output recording). These timestep increments can differ for differing events, and should be chosen for the appropriate level of fidelity. Timesteps chosen to be too coarse may not capture key behavior in the simulation (e.g., fast-moving network nodes may need finer timesteps to accurately capture small-scale fading effects, while relatively coarse timesteps would be fine for capturing large-scale fading).

TABLE 7-1. Continued

Design Choice Element	Description
Distributed or Single-Host	Generally, developing a complete network simulation to handle simulation tasks in a distributed manner among multiple host platforms will increase scalability of the simulation; however, complexity of the simulation increases as the required control message exchange to coordinate multiple host execution platforms increases. A single-host approach may be more suitable in cases where the simulation functionality does not require a substantial number of nodes to be simulated, or an increased fidelity of layer modeling.
In-house or off-the-shelf simulation components	Developing a complete network simulation from the ground up can be an expensive and time-consuming undertaking. As such, developers may choose to employ off-the-shelf simulation components that model parts of the network functionality, or even the entire network functionality; however, developers should be aware that off-the-shelf simulation implementations may not implement those functions the developer desires, or may not implement the functions with sufficient fidelity. If choosing an off-the-shelf component or platform, the developer is encouraged to fully understand the features, capabilities, and limitations of such a platform.
Set of outputs	The output set of the simulation should be designed so that it achieves the answer to the question posed to the designers. Generally, more outputs will require more message exchange via reporting from the simulation entities to the simulation engine; however, a designer may want to consider a selectable approach to the output set to enable more effective debugging, scalability, and flexibility.

TABLE 7-2. Complete Network M&S Platforms

Tool
OPNET
http://www.opnet.com
QualNet
http://www.qualnet.com
NS2
http://www.isi.edu/nsnam/ns

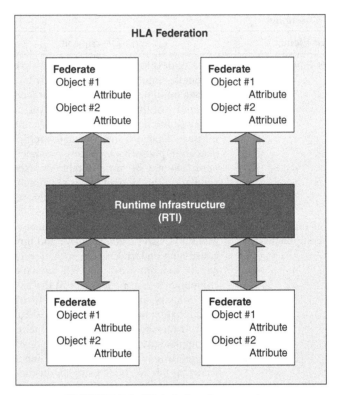

FIGURE 7-2. HLA federation overview.

federation control during the course of initialization and execution. The RTI is a key element within HLA and is essential to compliance with HLA. The internal functions and federate services it provides are abstracted enough to foster modular design and enable scalability by allowing additional federates to be added as the federation goals change. In a typical implementation, the RTI is likely to be a host computer or a cluster of such computers, maintained separately from the rest of the federation. Processing requirements for the RTI will vary substantially between federations, depending upon the number of federates, their interactions and data exchange requirements, timestep configuration of the federation, and required inputs and outputs of the federation during initialization and execution. Designers and developers responsible for a full HLA federation design will need to take into account these issues to achieve the federation goals while balancing the requirements for runtime execution speed, scalability (if needed), and modularity.

Prior to adoption by IEEE, the set of standards defining HLA was sponsored by the U.S. Defense Modeling and Simulation Office (DMSO). HLA version 1.3 [130–132] was the final version of the HLA standard while under

TABLE 7-3. HLA Standards Documents

Standard Title	Description
IEEE Standard 1516-2000, IEEE Standard for Modeling and Simulation (M&S) High Level Architecture (HLA)—Framework and Rules [74]	Overview of the HLA
IEEE Standard 1516.1-2000, IEEE Standard for Modeling and Simulation (M&S) High Level Architecture (HLA)—Federate Interface Specification [75]	Interface requirements and general specification for federates
IEEE Standard 1516.2-2000, IEEE Standard for Modeling and Simulation (M&S) High Level Architecture (HLA)—Object Model Template (OMT) Specification [76]	Format and syntax for information in HLA object models

U.S. DMSO sponsorship. For the IEEE, there are three documents that describe the components of HLA. These documents are summarized in Table 7-3.

Also of note to the designer may be the IEEE Recommended Practice for HLA Federation Development and Execution Process (FEDEP) [77]. This process details a set of common steps that developers, users, and sponsors may follow to develop successful federations. More detail on this process is provided latex in the chapter.

The remainder of this section will provide an overview of the HLA, based on the references provided above. There are three primary components to the HLA:

1. Rules. Ten rules have been defined to describe the general principles that define the HLA.
2. Interface specification. The interface specification contains a description of the functional interface between different simulations, known as "federates," and the HLA RTI.
3. HLA Object Model Template (OMT). The OMT is a specification of common formats and structures for the documentation of HLA object models.

HLA enables distributed simulation systems (physically or logically) to be unified in a single simulation environment. Such an environment is referred to as a HLA federation. Each individual separate simulation entity within the federation is referred to as a federate. Oftentimes the HLA framework documents will refer to "objects" within federates. Such objects could include typical simulation entities like a logical instantiation of a network node or a network router. In this sense the object-oriented view is similar to that employed in typical coding practice today—but HLA itself does not comply with all of the rules that govern object-oriented design. HLA is intended to be more high level and provide a framework to guide designers and developers in providing a common simulation environment to disparate simulation systems.

7.2.1 HLA Rules

This section discusses the 10 rules that are delineated in the HLA standard (IEEE 1516.1). The rules are defined to provide a standard structure adhered to by HLA-compliant federations. The rules are defined in [75] as follows:

1. Federations shall have an HLA Federation Object Model (FOM), documented in accordance with the HLA OMT. The FOM is essential because it contains the agreement reached between federates on how data will be exchanged using HLA services during the federation execution process. HLA itself does not dictate which data are included for exchange within the FOM, because this is a choice for the designers and developers of the federation architecture. Generally, more data exchange between federations will be necessary for higher fidelity, depending on the desired functionality for the simulation; however, an increased data exchange load between the federates may levy additional requirements on the entire federation as a whole to maintain runtime execution goals.

2. In a federation, all representation of objects in the FOM shall be in the federates, not in the RTI. By keeping the object representations and the ability to change properties of the objects within the federates themselves, this keeps the RTI functionality discretely separate; however, the RTI may access such properties in a read-only fashion to garner information to support its own services.

3. During a federation execution, all exchange of FOM data among federates shall occur via the RTI. By requiring the RTI to act as the delivery and handling agent for all FOM data exchange between the federations, this ensures maximum chances to maintain sufficient control over the federation as a whole while maintaining interface integrity. Otherwise, if federates were allowed to exchange FOM data directly to each other, overall simulation control and synchronization may be much more difficult to maintain.

4. During a federation execution, federates shall interact with the RTI in accordance with the HLA interface specification. This requires that a standardized interface be defined and adhered to based on the HLA interface specification and ensures that federates communicate in a standardized way to the RTI.

5. During a federation execution, an attribute of an instance of an object shall be owned by only one federate at any given time. This prevents multiple attributes of a single instance to be existent in the federation at any point in time. It prevents inadvertent object duplication and maintains integrity of the data (i.e., allowing two objects to represent the same simulated system in two different federates at the same time).

6. Federates shall have an HLA Simulation Object Model (SOM), documented in accordance with the HLA OMT. The SOM includes all the object classes, attributes of the classes, and interaction classes of each federate that are allowed to be made public within the federation.

7. Federates shall be able to update and/or reflect any attributes of objects in their SOM and send and/or receive SOM object interactions externally, as specified in their SOM. Each federate within the federation can readily exchange information from object instantiations to others within the federation if required.

8. Federates shall be able to transfer and/or accept ownership of an attribute dynamically during a federation execution, as specified in their SOM. This enables federates to pass object ownership if needed, and allows for maintaining Rule #5.

9. Federates shall be able to vary the conditions under which they provide updates of attributes of objects, as specified in their SOM. This rule allows federates to set thresholds that could change dynamically to enable when (or when not) to update attributes of their objects and make those attributes available to others via the RTI.

10. Federates shall be able to manage local time in a way that will allow them to coordinate data exchange with other members of a federation. To foster synchronization among federates, data exchange is generally recommended to occur at times that are well known to all federates. By managing the local time appropriately, federates will be more likely to maintain a sufficient level of synchronization, which is essential to a successful federation execution.

These 10 rules form the basis of creating an HLA-compliant federation. The intent of the rules is not to limit the designer/developer, but rather to provide a framework that adheres to the primary goal of HLA: to abstract the behavior of disparate federate modeling and simulation systems and foster successful interoperability between those systems to achieve a scalable, expandable, flexible federated simulation. However, it is important to note that implementation of HLA does not come without cost—the designer/developer may find that abstraction of the problem into object-oriented frameworks like HLA come at the expense of runtime, development cost, and complexity.

7.2.2 Interface Specification

The interface specification contains a description of the functional interface between different simulations, known as "federates," and the HLA RTI. The standard in [75] is over 475 pages and contains a very detailed interface specification for HLA. The purpose of this section is to provide an overview of the interface specification. If the designer/developer is interested in implementing full HLA, [75] will provide a more detailed treatment.

There are seven service groups that form the basis of the HLA interface specification. These service groups are divided in accordance with the basic functions of a federation. The service groups are as follows:

1. Federation management
2. Declaration management
3. Object management
4. Ownership management
5. Time management
6. Data distribution management
7. Support services

Together, these service groups describe the necessary interfaces between the RTI and the federates. Reference [75] states that *any* software that maintains the role of the RTI shall implement all of these services, so long as they are initiated by a federate. Consequently, any software that maintains the role of a HLA federate shall implement all of these services, so long as they are initiated by the RTI. The next seven subsections contain a more detailed description of each of the service groups and the purposes served by each.

7.2.2.1 Federation Management   The "Federation Management" service group is responsible for functions relating to controlling, creating, modifying, and deleting federation executions. Figure 7-3 illustrates the basic states of a federation execution [taken from reference 75].

Here, a federation execution state is shown to either exist or not exist. Basic functions such as joining the federation execution or removing (resigning) from the federation execution are shown for the federates. A federation execution is created or destroyed to move between the "exist" state and the "not exist" state. Federates may only join the federation once the federation execution is existent. Federates generally perform the exchanges to join a federation (via the RTI) as indicated in Figure 7-4.

Here, for a federate to join the federation execution, it must send a request to the RTI. Then, the RTI and the federate will establish initial data requirements through bi-directional exchange if needed. Once the normal federation execution process has started, features like advancing time in

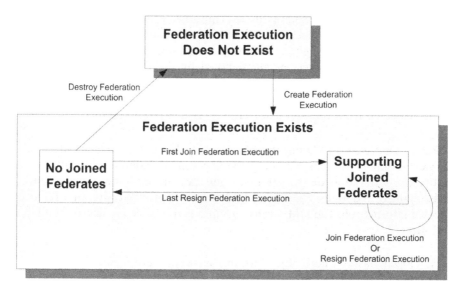

FIGURE 7-3. Federation management overview.

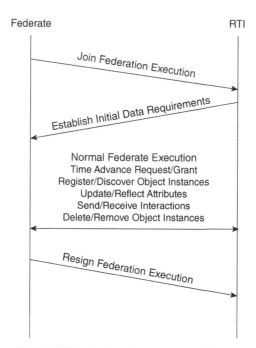

FIGURE 7-4. Federation joining exchange.

the federate, registering and/or discovering instances of objects within the federate, updating or reflecting attribute values within the objects, sending/ receiving data, or deleting/removing object instances are functions that may be performed. These functions are handled by the federate through direct exchange with the RTI. Once the federate is finished performing functions for the federation, it may resign execution from the federation and at that point is disconnected from the RTI and removed as a federation entity.

7.2.2.2 Declaration Management   The "Declaration Management" (DM) service group is responsible for providing joined federates the ability to notify the federation of the intent to generate information. Consequently, these joined federates must use DM services provided by the RTI to receive information. The DM service group is responsible for determining the following:

1. Object classes where object instances are registered
2. Object classes where object instances are discovered
3. Instance attributes able to be reflected or updated
4. Interactions that may be sent from a federate
5. Interaction classes used to receive interactions
6. Available parameters to be sent/received

7.2.2.3 Object Management   The "Object Management" service group handles registering, modifying, and deleting object instances. Furthermore, the service group handles sending and receipt of interactions.

7.2.2.4 Ownership Management   The "Ownership Management" service group is used by joined federates and the RTI to transfer the ownership of any instance attributes that exist among the joined federates. The ability to perform this function is required to support cooperation among the federates to model a given object instance across an entire federation.

7.2.2.5 Time Management   The "Time Management" service group is used to enable a federation to set the order of the delivery of messages throughout its execution. By using the functions provided in this service group, messages are sent and received by different federates in a specific order that maintains simulation coherency.

7.2.2.6 Data Distribution Management   The "Data Distribution Management" service group is used by joined federates to reduce the frequency of transmitted and received data that is deemed irrelevant. This particular service group has the goal of reducing the unnecessary overhead of message exchanges that may contain more information than is needed.

7.2.3 HLA OMT

The OMT is a specification of common formats and structures for the documentation of HLA object models. Reference [76] is the complete specification for the OMT. It includes the methodologies to define objects, attributes of the objects, inter- and intra-federate interactions, and parameters. Remember that a typical HLA federate is a simulation that provides certain functions to the federation as a whole. The OMT associated with this federate will describe the formats, structures, objects, attributes, interactions with other federate simulations, and any parameters that exist to define the federate and the objects within its responsibility. In this sense, the OMT is an abstraction of the federate. Any designer or developer responsible for developing an HLA-compliant federation will need to consider an OMT in the context of each of the separate federate simulations and in the context of the entire federation. Reference [76] maintains a high-level view of these definitions—that is, it is not specific to any particular application (e.g., a simulation that aims to model a wireless network). This section provides an overview of the HLA OMT as it is defined in [76].

The motivation behind defining an OMT in a formal way is considered important because it helps to:

- Provide a common methodology for the specification of data exchange and fosters more coherent coordination amongst the federation entities
- Provide a common methodology for capability descriptions for potential federation entities
- Foster common tool set designs and applications to further develop HLA object models as they evolve or change

An OMT may be scoped to describe an individual federate (known a simulation object model), or a set of federates that form a federation (known as a federation object model). The OMT provides the necessary processes to define the interfaces between simulations to foster reuse of components and interoperability between disparate federates.

7.2.3.1 *OMTs for the Individual Federate: SOMs* When defining an OMT for an individual federate, the OMT is known as a SOM. An SOM will contain key information about the federate functions and capabilities it provides to the federation, including what information may be provided and exchanged between the federate and the federation. The standardized format for the OMT, when implemented properly in the federation design, enables an apples-to-apples comparison between federate candidates and their functions to determine suitability within the context of the federation objectives.

7.2.3.2 *OMTs for the Entire Federation: FOMs* When defining an OMT for an entire federation, the OMT is known as a FOM. A FOM will contain the common set of required communications exchanges between the federates. Classes that define objects and the interactions between them within the

federates are listed, along with attributes and parameters that describe the objects and interactions. Each of the components of the FOM provide the common basis for the required interoperability between the federates; however, the FOM itself does not provide the functions that achieve this interoperability—the implementation process in the HLA federation design will achieve this. The FOM is a guide that helps designers and developers define, in a common format, the required interactions, so that the implementation process can be achieved more effectively.

7.2.3.3 Components of the OMT   Any FOM or SOM will contain particular component tables identified in [76]. This section provides a brief description of each of the component tables required. The reader is encouraged to consider reference [76] for more detail. The component tables are summarized in Table 7-4.

Following the OMT model as defined in [76] will provide designers and developers with a modular, abstracted definition of federate interactions, objects, attributes, and parameters. The OMT is intended as a guide to assist designers and developers by fostering interoperability between federates and the federation and provide key documentation relating to the relevant federation entity functions. It does not provide the steps for implementation, however. The implementation phase of federation construction will use the associated OMT FOM and SOMs as guidelines to build federates and the associated RTI functions that are necessary to foster data exchange and interoperability between the federates.

7.2.4 FEDEP Model

The FEDEP model is defined in the HLA standard [77] as a seven-step process that should be followed by all federations for development and execution. The scope of the FEDEP is to define the processes and procedures recommended for users of HLA in order to properly develop and execute different federations. The standard is largely viewed as a high level framework with substantial flexibility for designers to employ HLA in conjunction with their own practices (whether company-driven or developer-driven). Figure 7-5 illustrates this process.

Here, the process is divided into seven distinct steps. The steps are as follows:

1. Define the federation objectives. Here, the users, sponsors, and development team work to define and agree on objectives for the federation, and document the necessary steps to achieve the objectives. Such examples of objectives could include learning more about a particular protocol interaction in the presence of a large-scale network environment, or determining whether a set of protocols is sufficiently interoperable to support backward compatibility.

TABLE 7-4. OMT Components [76]

Component Table	Description
Object model identification	Lists the pertinent identifying information within the HLA object model. Such information could include the appropriate points-of-contact for the object model design, or an overview of the purpose of a particular object, parameter, or federate. This information is likely to be employed in developing future object models for other HLA federations and is intended to help foster re-use of the content provided in the current object model.
Object class structure	An enumeration of all of the federate or federation classes that define the objects, along with any sub-classes that could be part of these definitions, and the relationships between each. This is similar to typical best-practices in software design for documenting classes and relationships between them.
Interaction class structure	An enumeration of all of the federate or federation classes that define any interactions between federates, and describe any sub-classes and associated relationships that are part of these definitions.
Attribute	An enumeration of the features of object attributes within a federate or federation. Attributes of objects are typically defined to represent particular simulated entities (e.g., a simulated router may have an attribute such as its IP address).
Parameter	An enumeration of the features of the interaction parameters that exist within a federate or federation. An interaction generally specifies how objects within federates exchange information to support a simulation function. Parameters are defined to clarify what information is exchanged for a particular interaction (e.g., a simulated router object may want to interact with another simulated router object by sending a routing update message—the interaction class defines this interaction explicitly, while the parameter table defines what information will be provided in the message from the source to the destination router).

TABLE 7-4. Continued

Component Table	Description
Dimension	An enumeration of the dimensions used for filtering any instance attributes and interactions. This table assists with limiting information exchanges to the appropriate scope defined for a federate and its associated interaction with another federate or to handle data exchange with respect to particular attributes. It can provide a scalable way of fostering appropriate levels of data exchange for these elements.
Time representation	Describes the time value representations for the federation or federate. Because federates and federations often keep track of time during an execution and events are triggered and correlated to outputs, this table is important to define the associated data types and representations for time so that it is clear how federates or federations will handle stepping through simulation time and exchanging information on timestamps.
User-supplied tag	Any tags that are used in HLA services are defined here. Federates can supply their own tags to associate with particular HLA services provided by the RTI. This table documents the federation agreements with respect to the datatype used with the tags.
Synchronization	An enumeration of the datatypes and their representation used in synchronization services. The RTI will provide synchronization services to the federates. These synchronization services generally provide simulation coherency as the federation execution takes place. The particular synchronization feature provided by the RTI to the federates is known as the Synchronization Points feature. For the SOM, a description of which of the synchronization points federates can adhere to is provided. For the FOM, a description of the agreed synchronization points for the federation is provided.
Transportation type	Describes the transportation mechanisms used. The RTI will provide various ways of transporting data via interaction and object instance attribute value changes. The type of transportation supported for these interactions (i.e., one-way, bi-directional, read-only) are described for federates in the SOM and are listed as a complete set of transportation supportable functions in the FOM.

TABLE 7-4. Continued

Component Table	Description
Switches	Specifies initial parameter settings for the RTI. Such parameters could indicate whether or not the RTI should perform certain functions on behalf of federates—such as updating federates upon discovering newly created objects or object attribute updates.
Datatype	Enumerates the data representation details for the object model. Any datatypes that are defined in the object, interaction, attribute, or parameter tables can be further described in this table.
Notes	Provides additional detail of any of the OMT tables or their items.
FOM/SOM lexicon	Provides definitions for all of the objects, attributes, parameters, and interactions used. This table fosters a more complete understanding of the purpose of these entities, potentially enabling reuse and assisting in maintaining interoperability.

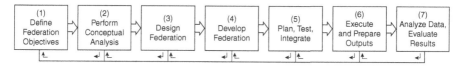

Iterative Development, Corrective Actions

FIGURE 7-5. FEDEP model overview.

2. Perform conceptual analysis. Depending on the problem the federation is attempting to solve, an analysis of how to develop the federation to approximate the simulated system behavior in the "real world" is conducted.

3. Design federation. Any existing federates that can be reused are identified, and adding new or modifying existing federates is conducted. Required functions are incorporated into federates, and a system design plan is created to foster federation development/implementation.

4. Develop federation. The FOM is created, agreements between federates are created, and modifications to existing federates and/or new federates are created.

5. Plan, integrate, and test federation. Required federation integration steps are executed, and testing is performed to determine if interoperability requirements are met.
6. Execute federation and prepare outputs. The federation is executed and its output data are pre-processed, if necessary.
7. Analyze data and evaluate results. Output data from the federation are analyzed and evaluated and results are provided to users and/or sponsors.

While these seven steps in the FEDEP model are common to all HLA federations, it is important to note that the amount of time to develop a federation will vary greatly depending on the particular problem the federation is trying to solve, the funding available to the developers of the federation, the requirements derived from the sponsors, the desired outputs and fidelity of the overall federation, the number of entities within the federation that are required to interact, and many other factors. Hence, the standard maintains a high level overview of the best practices and procedures to achieve a successful federation with the desired features, but does not necessarily address step-by-step details on how to achieve a fully functional federated simulation for any given problem. The sub-sections following describe the FEDEP model from a more detailed perspective as a reference to the designer who may employ this process. Figure 7-6 illustrates the in-depth model.

Figure 7-6 illustrates the primary activities that must take place within the federation development and execution. It is intended only as a starting point and does not preclude any particular design flow agreed upon by the design team. Each task is intended to be customized to the needs of a particular design. Furthermore, much of the tasking in the figure is intended to be conducted in parallel—the order is not necessarily strict.

The next seven subsections cover in more detail the FEDEP steps from [77]. For a designer or developer who is interested in following the HLA standard in developing a federated simulation, these steps may be very useful.

7.2.4.1 *Step 1: Define the Federation Objectives*   In any good simulation design, it is important to define what objectives the simulation is trying to achieve. For HLA-compliant federations, this is the first step. In reference [77], there are two key steps to define the federation objectives: identify the user and/or sponsor needs, and develop the objectives. The steps are provided in more detail below:

1A. Identify the user and/or sponsor needs. Here, it is important to note that the users of the simulation or the sponsor who is funding the simulation development effort are trying to solve a problem that the simulation is intended to solve. It is important to identify what needs these communities have to make sure that the designers and developers

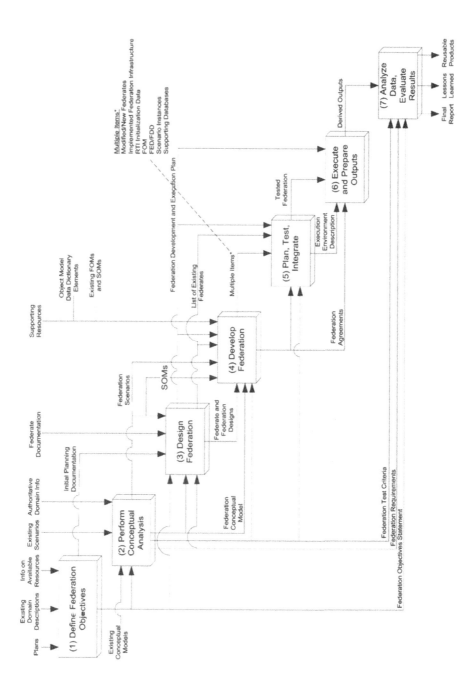

FIGURE 7-6. FEDEP model detail.

161

responsible for the federation development keep these needs in mind and achieve the goals of the users and/or sponsors. Items that should be addressed here include the descriptions of the systems of interest to be modeled within the federation, the resources available to support the federation development process, and constraints that may limit how the federation is developed. More detail provided at this stage will give all federation stakeholders a clear understanding of expectations, resources, and goals and minimize communication ambiguities later in the process.

1B. Develop objectives. Here, refinement of the needs identified in step 1A will take place, to develop a more specific set of objectives for the federation. An objectives statement is recommended by [77] to provide the basis for generation of requirements. Developing more specific, measurable objectives enables the federation development team to assess progress towards achievement of the federation goals and the higher-level needs delineated in step 1A. Assessments of feasibility and risk should also occur at this step. Such assessments would include the issue of selecting particular tools to assist or form the components of the federation design. Deciding which available tools to use should take into account cost, applicability to the federation objectives, compatibility with the federation architecture, and personal preferences of the users, sponsor and/or designers of the federation.

7.2.4.2   Step 2: Perform Conceptual Analysis   In this step, the federation stakeholders are conducting an in-depth analysis to determine how to represent the behavior they seek to reproduce in the federation. Generally, such behavior is an approximation to the "real world" and as such the designers and developers of the federation need to determine how to represent the behavior within the federated simulation. Usually such behavior will be driven by specific scenarios or operating conditions in the real world. Definition of the appropriate scenarios is also essential in this step. There are three key steps here: develop the scenarios of interest, develop a federation conceptual model, and develop the federation requirements. The steps are provided in more detail below:

2A. Develop the scenarios. Here, clear functional descriptions of the scenarios expected to form the basis of the federation behavior are developed. Any constraints on the federation should be considered to shape scenario development. Oftentimes, developers and designers may find a variety of already-defined scenarios from other federations or simulations that may be useful to leverage. Generally, a federation scenario includes a description of the number of entities within the federation and the types of each entity, along with their respective capabilities, relationships, and behaviors, and how these may change over time. Furthermore, environmental conditions are generally key defining elements for a scenario description. These can include the physical location of the entities, the weather conditions, or the terrain, for example.

2B. Develop the federation conceptual model. The developers and designers of the federation develop a conceptual representation of the problem space posed by the user needs and the more-specific federation objectives. Here, the model is used more as a representative guide that contains assumptions, limitations, and behavioral and functional descriptions of federation entities. The model can be used to assist in defining scenarios for the federation, and can be used to identify problems that may exist early on in the federation development process. It is important to carefully consider the conceptual model as it will form the guiding principles for the designers and developers during the design phase of the federation.

2C. Develop federation requirements. During step 2B, it is likely that requirements will be defined for the federation. The requirements are generally based on the objectives defined in Step 1B, and should be sufficiently defined to provide clear guidance for implementation while maintaining testability to verify the requirement has been met. Federation requirements should also consider execution needs of federation users, including control, monitoring, and logging. The necessary fidelity of the federation will be defined here as well, and will likely impact selection of federates.

7.2.4.3 *Step 3: Design Federation*

In this step, the federation design is solidified. Federates are identified that are suitable for the federation. New federates may be created, or existing federates may be re-used or adapted for use within the federation. A plan for developing and implementing the federation is delineated. There are three key steps here: select the federates, prepare the federation design, and prepare a development plan. The steps are provided in more detail below:

3A. Select federates. Individual simulation components are considered in this step as potential federates. This may involve using off-the-shelf components that could be employed within the federation architecture (so long as those components contain the necessary interfaces to comply with the federation interfaces) or developing new simulation components that achieve the federation objectives. Furthermore, existing federates could be modified or employed without modification as part of the federation. The designers and developers are encouraged at this stage in the process to utilize all information resources available on current simulation components (whether off-the-shelf or in-house designed) to make the correct decision on whether to select a potential federate for inclusion in the federation. By reviewing such available information (i.e., design documentation) the designers and developers will likely achieve a more clear understanding of whether a particular simulation component meets a set of the federation objectives or not. Also, the choice of particular simulation components may limit overall federation functionality—so designers and developers should not consider any choice only in the context of a single desired function, but also consider that choice in the context of the overall desired federation goals.

3B. Prepare federation design. Once federates have been chosen, the next step is to prepare an appropriate federation design. In this step, federates are assigned responsibility for entities and actions defined in the federation conceptual model. Assessment of whether the federates can effectively represent these entities and actions is conducted. Trade-off studies may be conducted here to determine the feasibility of using the selected federates (i.e., security, time management, federation management, runtime performance, and scalability).

3C. Prepare development plan. A coordinated plan will be developed here to guide development, testing, and execution for the federation. Generally, it is important for federation stakeholders to work closely during this process and agree on a common set of federation goals, requirements, and to identify any methodologies needed. Furthermore, milestones may be identified for each federate in its lifecycle process development.

7.2.4.4 Step 4: Develop Federation   This step includes developing the FOM, modify any federates as needed, and finally prepare the federation for integration and testing. There are four steps here: develop the FOM, establish federation agreements, implement federation designs, and implement the federation infrastructure. The steps are provided in more detail below:

4A. Develop FOM. The FOM development process will support data exchanges required between the federates to meet the federation objectives. There are many different ways to develop an FOM, each of which has its own advantages and disadvantages. The designers and developers should consider which of the methods may be the most appropriate for the federation. Some of the steps listed in [77] are provided below:

- Develop a "clean-sheet" FOM by using the federation scenario and federation conceptual model while applying any existing standards.
- Merge together other SOMs of all the existing federates, removing any aspects of a particular SOM that do not apply.
- Start with the SOM that is closest to the desired FOM, remove aspects that do not apply, and merge parts of other SOMs as appropriate to fully represent the federation.

- Use a FOM from a previous, similar application and modify as appropriate.
- Use a FOM that provides a common reference frame for a given user community. Remove elements that are not appropriate.
- Consider reusable OM components to develop the FOM.

4B. Establish federation agreements. Operational agreements are required amongst the designers and developers of the federation and management that are not documented within the FOM. Federation developers and designers need to consider what agreements are required in addition to the FOM and how these agreements are documented. For instance, licensing agreements for off-the-shelf simulation components employed as federates are examples of such agreements. External agreements such as these may limit the development and/or execution of a particular federation, and need to be considered in the context of the federation project plan.

4C. Implement federate designs. Modifications to federates are implemented in this step to ensure that the federates can represent objects and behaviors assigned to them within the federation conceptual model. Furthermore, the federates may be modified to ensure they can support the data exchange requirements to and from other federates as defined in the FOM.

4D. Implement federation infrastructure. This step includes the initialization, configuration, and implementation of the required infrastructure to support the federation and verify its abilities to execute and provide communications capabilities to all the components within the federation. The federation network design is developed here, along with any installation and configuration of supporting hardware. Preparation of the hosting facilities for the federation is conducted.

7.2.4.5   *Step 5: Plan, Integrate, and Test Federation*   This step involves finalizing the federation execution planning, connecting federates, and testing the federation. Three steps are key here: planning the execution, integrating the federation, and testing the federation. The steps are provided in more detail below:

5A. Planning the execution. During this step, a description of the execution environment for the federation is created, along with an execution plan. Federate and federation performance requirements are delineated. If security is a consideration, development of a security test and evaluation plan is conducted. Operational planning is also considered key during this step. Such planning needs to consider the required operational and support staff during an execution run and what steps should be taken to prepare the federation for execution prior to a run. Starting, executing, and concluding steps should be delineated for each run.

5B. Integrate federation. Here, all of the federates and other federation participants will be unified together into a single operating environment. This does not necessarily require that all members of the federation operate with the same operating system—rather, it requires that all of the components are properly installed and interconnected so that the federation goals are achieved. In practice, it is very likely that federates will only comply with federation-defined interfaces in the FOM; their host platforms may be vastly different for a single federation. It is expected that integration of the federation will be closely tied to the testing of the federation because it is during this phase that problems integrating equipment are found and testing procedures become more refined.

5C. Test federation. Here, all the federation members are tested within the federation to ensure interoperability so that they may achieve the federation objectives. There are three levels of testing defined for the HLA:

a. Federate testing: Each federate is separately tested to make sure software is correctly implemented with respect to the federation requirements.

b. Integration testing: The federation is tested in its entirety to verify basic levels of interoperable components.

c. Federation testing: The federation is tested in its entirety to verify it can achieve the federation objectives. This is generally a more rigorous test than the prior step.

7.2.4.6   *Step 6: Execute Federation and Prepare Outputs*   Here, the federation is executed and outputs are pre-processed as appropriate. There are two key steps: executing the federation and preparing the federation outputs. These steps are summarized in more detail below:

6A. Execute the federation. All of the federate members are activated within the federation for formal execution—this is where the rubber meets the road. Management of the execution process and storing/collection of data are both critical to a successful federation execution. Controlling and monitoring the execution during runtime is essential; such monitoring metrics could include CPU usage or network loading statistics. During the data collection process, the desired set of outputs are stored in an appropriate place for retrieval later. Any secure federation executions may require additional steps for each run, to ensure that all federates and procedures are certified.

6B. Prepare federation outputs. Any pre-processing of the data collected in Step 6A may occur here. Generally, pre-processing involves arrangement of the data in a form that fosters a clear way to formally analyze the data.

7.2.4.7   Step 7: Analyze Data and Evaluate Results   Once a federation execution is complete, data generated during the execution need to be analyzed and evaluated to determine the answer to the problem the federation and the specific run have been designed to solve, or at the very least, yield insight. Reporting of the results is provided to the user and/or sponsor. There are two key steps here: analyze the data, and evaluate and feedback the results. These steps are described in more detail below:

7A. Analyze the data. Here, any outputs provided by the federation execution are analyzed—data may be provided through a variety of ways, and tools and methods will be used here that are deemed appropriate for the analysis. Such tools could include off-the-shelf software packages (such as Microsoft Excel) or specific software tools designed especially for the federation.

7B. Evaluate and feedback the results. Here, the results are evaluated to determine whether they have met the federation objectives. This includes determining whether the current set of results meets the federation test criteria findings. Any reusable federation products from the execution are stored in an archive as appropriate for reuse in other HLA federations or for the domain of interest.

The seven steps described in this section should provide the designer/developer with an appropriate process to design HLA-compliant federations. While the HLA system itself can be very overhead intensive if employed in a strict fashion, it does allow for flexibility in the design of a particular federation. In this sense, the HLA method is widely applicable to a variety of simulation designs. Furthermore, the steps outlined have been derived from years of experience in designing HLA-compliant federations. The HLA process itself is useful to any designer or developer who requires interoperable components for a simulation. It need not be employed to its fullest extent, but some of the general principles provided here may assist the designer or developer in creating a more interoperable, sustainable, and successful simulation.

7.2.5   Example Federation: Complete Network Model

The purpose of this section is to provide the designer/developer with an example of an HLA federation at a conceptual level for a complete network simulation. Here, an overview of the federation will be provided, with a list of the federates used, their object functions, and specific attributes and interactions that could occur between the federates. Figure 7-7 provides an example of the HLA federation methodology applied to a wireless mobile network simulation.

Consider a complete wireless mobile network simulation federation that consists of four primary federates:

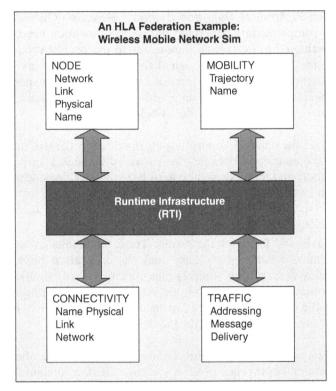

FIGURE 7-7. HLA federation example: wireless mobile network simulation.

1. The NODE federate, which maintains the functions of network nodes within the federation execution
2. The MOBILITY federate, which maintains the functions of movement of the network nodes within the federation execution
3. The CONNECTIVITY federate, which maintains the network topology of the network nodes within the federation execution
4. The TRAFFIC federate, which handles the sending of user traffic for the simulation

7.2.5.1 Example Federate Decomposition   Within each of the federates, there are a set of objects, each with attributes. This section provides a decomposition of each of the federates in the following format:

- Federate NAME
  - Object #1
    - Attribute a) of Object #1
    - Attribute b) of Object #1
    - . . .

- ○ Object #2
  - ▪ Attribute a) of Object #2
  - ▪ ...
- ○ Object # ...
  - ▪ ...

The purpose of this section is to provide an overview of how network functions can be embedded within objects of a federation. Any designer or developer who chooses to implement the HLA federation method may employ different objects and attributes, depending upon which functions of network M&S are considered the most important for the task. It should be noted that the example provided here is at a very high level and is not intended to capture a particular implementation, but rather provide the reader with an overview of the basic decomposition of a federate in the context of network M&S.

For the NODE federate, the objects and their associated attributes are as follows:

- • NODE
  - ○ Network Layer
    - ▪ Protocol type (i.e., IP)
    - ▪ MTU size
  - ○ Link Layer
    - ▪ Access method (i.e., CSMA)
    - ▪ Link Layer Address
  - ○ Physical Layer
    - ▪ Radio frequency
  - ○ Name
    - ▪ Name of node

Here, the attributes listed are included for the physical, link, and network layers for the node. In this sense, the NODE federate only provides details on the communications stack for the given node, as well as a Name object which contains a "Name of node" attribute for identification purposes. Note that the "Name of node" is italicized. The italic indicates that the ownership of this attribute can be transferred among different federates. It will be clear later why this is important. The NODE federate performs the network, link, and physical layer functions. Here, that could include passing data between each of these layers in the communications protocol stack or modifying object attributes such as radio frequency (if the physical layer could change its frequency).

For the MOBILITY federate, the objects and their associated attributes are as follows:

- MOBILITY
  - Trajectory
    - Velocity
    - Starting position
    - Current position
    - Update rate
  - Name
    - Name of node

The MOBILITY federate is a simple way to represent the mobility of nodes within the network simulation federation. Here, the objects defined for mobility are the Trajectory object and the Name object. The trajectory object contains four attributes: velocity, starting position, current position, and update rate. The Name object contains one attribute: Name of node. Once again, the Name of node can be transferred among different federates. The federation would use the MOBILITY federate to handle node movement. Each node in the simulation could be associated with a respective MOBILITY object that would contain its velocity, starting position and current position (updated by the MOBILITY federate, at the specified update rate). The MOBILITY federate would provide the function of updating the position of the objects at the specified rate.

For the CONNECTIVITY federate, the objects and their associated attributes are as follows:

- CONNECTIVITY
  - Name
    - Name of node
  - Physical Layer
    - Names
  - Link Layer
    - Names
  - Network Layer
    - Names

The CONNECTIVITY federate handles determining the connectivity of each of the layers of the protocol stack simulated in the federation. Here, there are four objects: Name, Physical Layer, Link Layer, and Network Layer. The Name object contains the "Name of node" attribute whose ownership can be transferred. The physical, link, and network layer objects each have a single attribute, "Names," which contains the names of the other nodes connected to the current "Name of node."

For the TRAFFIC federate, the objects and their associated attributes are as follows:

- TRAFFIC
  - Addressing
    - Source name of node
    - Destination name of node
  - Message
    - Size
    - Timestamp
  - Delivery
    - Sent successfully
    - Received successfully

Here, the TRAFFIC federate contains three objects: the Addressing object and the Message object. The Addressing object contains two attributes: the source and destination node names. The Message object contains two attributes: the size of the message and the timestamp of the message. The Delivery object contains two attributes: sent successfully and received successfully. It is expected that the TRAFFIC federate would be used as the basis for the traffic profiles for a given simulation execution. That is, every message to be sent in the federation would be an instantiated object within the TRAFFIC federate, which would keep track of message attributes like size and timestamp and trigger node "transmissions" in the federation. Furthermore, the Delivery object would enable the simulation to keep track of whether each message object was sent and received successfully (or not).

For a typical federation execution using this example, it is expected that the following functions would occur:

1. Initialization of federation
   a. Instantiation of all NODE objects
   b. Initialiation of MOBILITY objects
      i. Starting points established for each node name
   c. Initialization of CONNECTIVITY objects and creation of starting point of topology connectivity
2. Execution of federation
   a. Send all message objects ordered by timestamp defined in the TRAFFIC federate from source to destination until first update trigger for MOBILITY and CONNECTIVITY federate functions
   b. Record success or failure of delivery
   c. Employ MOBILITY federate to update node locations based on trajectories

    d. Employ CONNECTIVITY federate to update node connectivity at the three layers

    e. Repeat steps a through d until all message objects have been sent

3. Reporting of results

    a. Output of message success rates

    b. Output of ending state of network connectivity and node locations

While not explicitly shown, it is clear that there are many functions that would be necessary to achieve a complete network simulation in this example. Message exchanges between each of the federates would need to occur on a regular basis during the federation execution because of the need to share information. The "Name of node" attribute will be shared among multiple federates in this example because it is expected that a complete instantiation of a node will need to be defined by its communications layers, connectivity, and position information. It is possible to implement such a function in another way—this case is only an example to illustrate to the reader how the HLA architecture entities map to functional entities defined for a wireless mobile network simulation.

7.3 COMPLETE NETWORK SIMULATION EXAMPLES

This section provides an overview of some network simulation examples that focus on providing fidelity at multiple layers of the simulation. Careful attention is paid to the specific features implemented at the various layers for each of the simulations. As always, any designer or developer will need to take into consideration their own needs and tailor a simulation design to specifically address those layers that are the most important to model within the context of the simulation goals.

7.3.1 Scalable Urban Network Simulation (SUNS)

The design of a simulation focused on determining the performance of wireless-based networks in tactical environments is detailed in [78]. The Scalable Urban Network Simulation (SUNS) is a combination of a full network layer simulation with a high-fidelity ray-tracing RF propagation model suite. The authors in [78] chose a high-fidelity, RF propagation model based on ray-tracing approaches to better model the urban environment where small-scale fading is a substantial concern, due to the constructive and destructive interference inherent in such environments. Figure 7-8 illustrates the SUNS concept. Here, an urban propagation database is pre-calculated and is interfaced to the OPNET simulation via an Application Program Interface (API). The SUNS client interfaces to the SUNS server, which is embedded within the OPNET simulation as a custom model to provide the server interface. The SUNS server

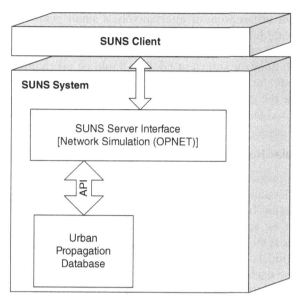

FIGURE 7-8. SUNS.

communicates with each of the network node instantiations within the OPNET model to enable, disable, establish, or relocate simulated communications, as directed by the SUNS client. Propagation loss is determined on a point-to-point basis by using the pre-calculated urban propagation database. The database can be generated using any urban propagation model, though the authors chose EMAG Technologies' EMLounge ray-tracing approach for their application.

In this example, it is clear the authors were specifically interested in modeling urban propagation and considering this aspect of the physical layer very important in modeling link effects in such an environment. The authors relied primarily on the OPNET modeler (and its associated limitations) to provide layer two through seven functionality while specifically requiring path loss calculations to be dependent on the SUNS client-server and associated urban propagation database. OPNET modeler can provide path losses using other models but the authors believed their approach provided improved fidelity over other options. To analyze wireless network performance in an urban propagation environment, this approach may likely be very powerful; however, this approach may not be suited to all applications. For instance, it is necessary to calculate path losses for ray-tracing models based on single source points—that is, for a given map of interest, the map must be divided into a grid of points—one point is chosen as the source, while the rest of the points are considered destinations. Each source-destination pair has an associated path loss calculation, and as such, the number of source-destination pairs can increase substantially for specific cases. Such cases include a large number of nodes within an urban area, or alternatively, a small number of nodes with each experiencing high mobility conditions. While the database is calculated prior to simulation execution, the time to populate the database based on the required calculations for sufficient fidelity should be taken into account. Ray-tracing methods for urban environments are generally considered high-fidelity—but at the cost of execution time compared with simplistic models such as lognormal fading models.

7.3.2   Vehicular Network Modeling

A generalized approach to modeling vehicular networks effectively is described in [79]. To understand the needs of a particular vehicular network model, a general abstraction between the vehicular motion of nodes and the network simulation can be considered. Figure 7-9 illustrates this abstraction. Here, vehicular traffic is modeled as a mobility model that is driven by driver behavior. Realistic models can be integrated where individual vehicular nodes are dependent upon each other's behavior (remember that we generally stop our car from running into the car ahead of us). Some vehicular network models in the literature take this behavior into account, while others use a simplified general approach based on velocity profiles independent of each user. Here,

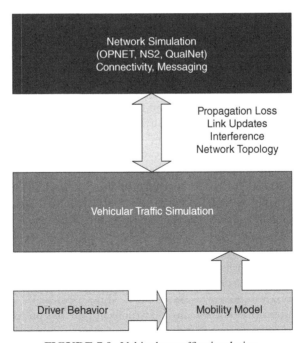

FIGURE 7-9. Vehicular traffic simulation.

the physical layer of the network is considered to be very important to the vehicular network simulation as a whole. As such, the traffic simulation is a key part of the overall complete network simulation. The traffic simulation will provide calculations that are key to determining achievable link rate between vehicular nodes by determining interference levels (caused from other vehicle nodes), propagation losses (including small- or large-scale fading), link updates (such as establishment of new links and tear-down of links no longer needed) and as such will heavily drive the network topology for the network simulation.

The authors provided an abstracted way to provide a vehicular traffic simulation based on some of the key elements that would comprise such a simulation. In this sense, their framework may be scaled or tailored to particular usage cases. As always, it is important to consider what degree of fidelity in modeling vehicular behavior will achieve the necessary degree of modeling detail for the overall network simulation. For instance, it may not be useful to develop a state-based model of vehicular motion where every vehicle's motion is dependent upon the surrounding vehicles if the simulation goal is to provide a statistical average result, such as average network throughput or average delay. However, if the simulation goal is specifically focused on determining performance of throughput and delay as a function of traffic patterns (such as rush hour versus non-rush periods), it may be very useful to implement a state-based approach since such results may not be determined with a sufficient degree of fidelity from a statistical analysis. Once again, designers and developers should consider the goals of their simulation and choose an appropriate method that models the necessary behavior to yield the appropriate insight. While this may sound like a simple task, it is often rife with trial-and-error approaches that bring the designer/developer of the simulation closer to the answer desired. In this sense, it is essential to peruse literature on the subject to see if others may have tried similar approaches.

7.3.3 IEEE 802.16e HLA Abstracted Simulation

The IEEE 802.16e standard is considered to be an increasingly relevant technology for providing metropolitan wireless networking capabilities. This technology is architecturally similar to cellular network technologies such as LTE, in that there are both base stations and subscriber stations, and the base stations are in control of the vast majority of communications. Furthermore, 802.16e allows subscriber stations to handoff between base stations. There are a number of features that can be modeled for a complete IEEE 802.16e simulation. This section aims to discuss some of these features for all five layers of the communications stack in Figure 7.1. IEEE 802.16e is a PHY and MAC layer specification, so any design that incorporates the five-layer stack will need to be developed with additional functionality above and beyond 802.16e. In particular, the network, transport, and application layer abstractions will need to be developed.

TABLE 7-5. IEEE 802.16e Mobile WiMAX Simulation Federate List

Federate Name and Description	Objects and Associated Functions
Host A high-level federate that represents a particular host platform that runs applications and has a network connection via an IEEE 802.16e-based subscriber station	**Application Layer** Responsible for providing application functionality (e.g., instant messaging or video chat), generating associated traffic streams, and monitoring performance of those traffic streams (number of packets dropped, retransmits). Inputs: application type and rate, bit streams from the transport layer federate. Outputs: bit streams to the transport layer federate. **Transport Layer** Responsible for maintaining TCP and/or UDP connections from the host to another remote platform. Maintains semantics and complies with key functions of these protocols (such as retransmissions for TCP and assigning port numbers for both TCP and UDP connections between hosts). Inputs: bit streams to and from the application layer federate, IP datagrams received from the network layer federate. Outputs: TCP or UDP segments sent to the network layer federate. **Network Layer** Responsible for IP address identifiers that uniquely identify a host. Inputs: TCP or UDP segments from the transport layer, IP datagrams from the 802.16 layer. Output: IP datagrams sent to the IEEE 802.16e federate, TCP or UDP segments sent to the transport layer federate. **IEEE 802.16 Layer** Responsible for modeling both the MAC and PHY of IEEE 802.16. In the case of a subscriber station, will handle the functions of authentication and association to a base station, as well as requesting resources to receive and transmit data. This federate also handles converting the MAC layer frames into bits at the PHY, in the form of the OFDM symbols as specified in the IEEE 802.16e standard. Inputs: subscriber station or base station, PHY configuration (transmit power, OFDM-specific elements), IP datagrams received from the network layer, and received complex baseband PHY signals from the channel federate. Outputs: complex baseband representation of the PHY signal sent to the channel federate via the RTI

TABLE 7-5. Continued

Federate Name and Description	Objects and Associated Functions
Base Station A high-level federate that represents a base station in an IEEE 802.16e-based network that is able to relay information between the two subscriber stations that are attached to the hosts in the example.	**IEEE 802.16 Layer** Responsible for modeling both the MAC and PHY of 802.16. In the case of a base station, some of the major functions include coordinating and assigning timeslots, assembling downlink and uplink maps, relaying data between subscribers via the channel federates, and assembling complex baseband PHY symbols from MAC layer frames as needed.
Channel A high-level federate that represents effects of the wireless channel on the signals (complex baseband PHY symbols) "transmitted" from the IEEE 802.16 federates.	**Propagation Effects** Propagation effects will include small-scale and large-scale fading effects, (e.g., based both on constructive/destructive interference concepts and lognormal path loss models, respectively). Inputs: propagation models. Outputs: changes in the signal representation based on the propagation model desired (usually an amplitude change only, but could affect frequency and/or phase as well). **Impulse Response** The impulse response federate could be used to operate on the complex baseband representation of the PHY symbols being generated from an IEEE 802.16 federate. Inputs: impulse response function H(t). Outputs: H(t) convolved with the PHY signal. Propagation model functions could be built into the impulse response function but some simulations may prefer a more simple approach by utilizing the propagation effects federate instead.

For the purposes of this example, it is assumed the simulation goals are to determine the performance of two application types (video chat and instant messaging) between two hosts that are utilizing an IEEE 802.16e network to connect to each other. Table 7-5 lists a notional polymorphic federation architecture for this simulation. It is assumed that typical internet protocols of today are used at both the transport and network layers (e.g., TCP, UDP, IP). Note that there are three primary federates in the simulation: the Host, the Base Station, and the Channel.

Figure 7-10 provides an abstracted overview of the HLA architecture for this example. Note that each federate connects to the RTI. Any control

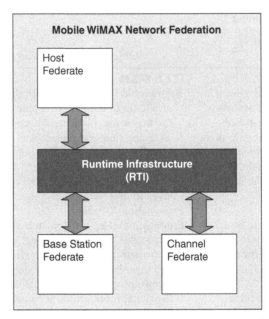

FIGURE 7-10. IEEE 802.16e Mobile WiMAX HLA federation.

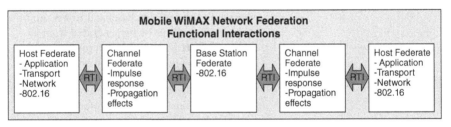

FIGURE 7-11. IEEE 802.16e Mobile WiMAX federation functional interactions.

functions are provided by the RTI, and it also provides the necessary interfaces for federate interactions.

Figure 7-11 provides an illustration of how these federates would functionally connect to each other in the simulation. Note that the RTI provides the interface between each of the federate instantiations. Also note that the objects listed within each federate block are handled exclusively by their parent federate.

This example has provided a very simple scenario—that of two hosts connecting to each other directly using video chat or instant messaging via an IEEE 802.16e-based mobile WiMAX network with a single base station. In practice, the designer or developer may wish to simulate a much larger number of hosts and/or base stations, and may look to explore the performance of other applications as well. However, the abstractions provided can be used as a basis for developing such a simulation. It is important to note there are no "right" or "wrong" abstractions—but any simulation must balance the choices for abstraction with the efficiency of the simulation in achieving the stated goals.

CHAPTER 8

Other Vital Aspects of Successful Network Modeling and Simulation

8.1 VERIFICATION AND VALIDATION

This is undoubtedly the most unpleasant aspect of modeling and simulation. And while certainly a burden, verification and validation (V&V) is a necessary step that cannot be avoided. Remember, the burden of proof is on the simulation developer! There are individuals that will remain skeptical of performance claims until it has been proven empirically. And this is somewhat deserved. Open literature is littered with claims of performance based on M&S results that prove non-existent once implemented.

Let us face an unpleasant reality: anyone can develop a network simulation. The fact that a simulation exists does not make it useful or trustworthy. What makes a simulation useful or useless is the care that goes into ensuring that a simulation not only models the correct aspects and behaviors of a system, but that it also does so with adequate fidelity. That is what V&V is all about. The establishment of a model's credibility via V&V is a necessity. In some cases, when a model or simulation is to be used to provide a critical input into a decision-making process, a formal accreditation process might also be required (referred to as verification, validation, and accreditation (VV&A). The VV&A process is illustrated in Figure 8-1.

Reference [80] defines verification as "the process of determining that a model implementation accurately represents the developer's conceptual description and specifications." Validation is defined as "the process of determining the degree to which a model is an accurate representation of the real world from the perspective of the intended uses of the model." There are seven steps in the V&V process: determine V&V requirements, initiate V&V planning, conceptual model V&V, design V&V, implementation V&V, application V&V, and acceptability assessment.

An Introduction to Network Modeling and Simulation for the Practicing Engineer, First Edition. Jack Burbank, William Kasch, Jon Ward.
© 2011 Institute of Electrical and Electronics Engineers. Published 2011 by John Wiley & Sons, Inc.

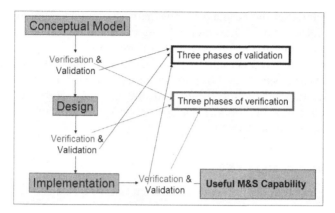

FIGURE 8-1. Verification and validation process.

8.1.1 Determine V&V Requirements

This step of the V&V process consists of defining requirements by which the success of the V&V effort will be judged. This step occurs after the method of M&S implementation has been selected (legacy, modify, or build new). These requirements include level of effort estimates, techniques to be employed, and logistic factors such as equipment needs, required manpower, and overall V&V cost.

8.1.2 Conceptual Model V&V

This step of the VV&A process ensures that the conceptual model adequately captures and meets the requirements set forth for the model. The conceptual model includes all equations, models, and algorithms that will be employed to meet the defined software requirements. The conceptual model also includes all assumptions and limitations that are to be made and how they may impact the ability of the model to satisfy its requirements.

Conceptual model validation ensures that the conceptual model provides an acceptable representation of the system of interest, and addresses fidelity requirements levied upon the model.

8.1.3 Design V&V

This step checks that the conceptual model is mapped to a design in an acceptable fashion. This is done to ensure that the model design accurately reflects the validated concept and associated requirements.

Design verification is very much like conceptual model verification. Once components within the conceptual model are mapped to software objects and functions, those objects and functions are reviewed to ensure that the mapping is complete and that the design captures the verified and validated conceptual model to meet software requirements. Design verification can be accomplished prior to the writing of a single line of program code.

Design validation refers to the analysis of the implemented software objects and functions to ensure that those objects and functions have been developed in a manner that adequately represents the intended corresponding component of the conceptual model. Design validation is typically accomplished through program code reviews.

8.1.4 Implementation V&V

This step checks that the implemented design (i.e., program code) matches the model design. This document addresses implementation verification, which is commonly referred to as *code verification*.

8.1.5 Putting the Steps Together into a V&V Process

In summary, there are at least three types of verification that must be considered: conceptual model verification, design verification, and implementation verification. Furthermore, there are at least three types of validation that must be considered: conceptual model validation, design validation, and implementation validation.

In addition, input V&V is often a very important part of the overall V&V process. Input V&V deals with the assurance that the inputs that are used to drive the simulation tool are valid and adhere to the correct formats. Output V&V deals with providing assurance that the output formats are correct and comprehendible in a way that they will not be misunderstood.

According to [81], "The verification process checks to see that the M&S implementation accurately represents the developer's descriptions and specifications. The subsequent validation applies the model to alternate data sets and attempts to determine the degree to which the M&S accurately reflects the real world."

The various components of the conceptual model are then translated into design components (intended software objects and functions), in the form of a software design specification. Design verification maps the various intended software components back to the corresponding components of the conceptual model. The various software components are then analyzed to ensure that the verified and validated ("V&Ved") conceptual model is completely captured within the software design. That software design, once implemented and verified, undergoes code verification. Code verification is intended to ensure that the various software elements have been implemented in a manner consistent with the developed software specification and retain the capability of the conceptual model. Code verification also ensures that the various individual software elements are interfaced together in a fashion such that the overall desired capability, based upon the V&Ved conceptual model and V&Ved design, is realized in the implemented overall software package.

Conceptual model and design validation ensures the accuracy of the representations created by the developer and the adequacy of those representations and mapped designs in meeting fidelity requirements. Implementation validation tests the V&Ved design, once implemented, to ensure that the final implementation provides sufficient fidelity to adequately serve its intended usage.

Unfortunately, this is a topic for which there are not many useful "how-to manuals." One helpful technique that is often employed in this process, particularly in validation of simulation results, is analysis of extreme conditions. Here, extreme input cases are utilized where the output is intuitively known a priori. If a WiMAX simulation is established and transmission powers are set to 0 Watts, is the resulting packet error rate 100%? If conditions are set up that should guarantee proper delivery of data, is the packet error rate 0%? These boundary conditions cannot be used to determine the accuracy of a simulation, but can serve well as a quick sanity check to determine whether errors are present.

While this process sounds quite formal and cumbersome, the level of formality should be a function of the importance of the M&S output. If M&S is to be used to generate a key input to a significant business decision, then great care should be paid to V&V. If M&S is to be used to assess the performance of a human safety-related system, then V&V should be more formal. If M&S is employed as an analysis tool to assist in an engineering trade study, then less emphasis can probably be placed on formality; however, these steps should occur in any M&S effort as they are mandated by proper engineering practices.

Another important point to make is that V&V never really ends. We have already seen that V&V is critical as the designer begins building a conceptual model up through the final implementation of the simulation; however, V&V should not stop once the simulation has been implemented. Rather, V&V should continue throughout the lifetime of the simulation (i.e., as long as the simulation is being employed). It is a common pitfall to claim that V&V has been completed only to find out that there have been errors present in the simulation that could have been corrected if diligence had been continued in the V&V process. Never trust the result from a simulation, even your own! The V&V process never ends.

8.1.6 Model Verification and Validation: A Practical View

So what does this all mean from a practical perspective? Below is a checklist that represents some of the key steps that should be taken to ensure a model's sufficiency:

1. What are the metrics of interest to be observed? Too often a simulation is developed without any formal articulation of the goals. This is a key

step in deciding the correct choice of simulation tool and of required detail to include in the model. For example, if the goal is to investigate the effects of link errors on routing protocol overhead, it would not be appropriate to employ network layer abstraction.

2. What is the required accuracy of the model? Too often simulation developers develop bit-true simulations when only high-level performance trends were of interest and a reasonable amount of error in any particular result is tolerable. This leads to a misuse of resources, focusing on simulation development when focus on data trend analysis might have been more appropriate.

3. Does the simulation design meet the desired goals? Here, a network simulation design has been articulated on paper, but not yet implemented. It is often possible to identify shortcomings in a simulation at this early stage in the development process. If the goal is to study the performance of TCP over an IEEE 802.11 wireless network, perhaps it is not the best decision to approximate IEEE 802.11 as a token ring network. In this simplistic case, rudimentary design review would identify this type of mistake.

4. Is the simulation implementation an accurate representation of the intended design? Here, steps like code reviews can be employed to ensure that the designed simulation has been faithfully implemented. If the designed simulation employed TCP Reno as a transport layer protocol, and standard TCP was actually implemented, this is an error in implementation that could be remedied through simple code review.

5. Does the simulation implementation sufficiently meet the original design goals? This question is the hardest to answer in the process. Sometimes there are results available in open literature that can be used as a point of comparison. Sometimes there are empirical data that can be used as a point of comparison. Sometimes extreme boundary condition analysis can be used as a point of comparison. This process consists of comparing simulation outputs with as many points of comparison as possible to build a "warm and fuzzy" belief that the simulation is producing trustworthy results.

This is obviously a simplified view of the process, and there are certainly many other ways to model the process. But the authors contend that if these questions are asked at the appropriate times during the development process and paid appropriate attention, then the liklihood of success increases dramatically. These questions, or any other framework in which you prefer to place the process, all attempt to force proper software engineering practices to become part of the simulation development process. It seems obvious that this would already happen; however, simulations are often considered differently from other types of software. This is not a piece of software that will be embedded in a device. This is not a piece of software that will act as a user interface for a larger system. This is a piece of software being developed as an engineering decision aid. Often times, less importance is placed on this latter application because it is not deemed as mission critical as some of the former examples; however, if engineering design decisions are to be based in part on simulation results, than sufficient rigor must be applied. The last thing anyone wants is for design errors to be introduced into a system because corners were cut in developing a computer simulation. For a more rigorous treatment of verification and validation, the reader is referred to some of the many excellent books on the subject [127–129].

8.2 DATA VISUALIZATION AND INTERPRETATION

The proper visualization of simulation outputs is critical for multiple reasons:

1. Proper understanding and interpretation of results
2. Determination of correct model execution

Particularly for the case of wireless networks that include mobility models, it is difficult to understand if the model is functioning correctly without a way to properly visualize the simulation results. Reference [53] suggests that proper visualization methods can assist in ensuring proper simulation execution. This is one of the key capabilities that the simulation designer obtains with using commercial simulation tools since these often include power visualization tools, sometimes providing the capability to show temporal views of the many aspects of the state of the network. Both OPNET and QualNet provide significant visualization capabilities. Historically this has been one of the greatest limitations of open source tools and undoubtedly the source of numerous implementation errors; however, the open source community has evolved in this area over the past decade. iNSpect is a NS-2 network visualization tool that can be used to provide both still and animated visualizations of network behavior in much the same way as commercial simulation packages. Developed by the Toilers research group at the Colorado School of Mines, iNSpect has been successfully tested on Ubuntu, Debian, Redhat, Fedora Core, and Windows XP (SP2). Unfortunately, as of the writing of this book, iNSpect does not compile in Cygwin. For additional information, the reader is referred to [82].

CHAPTER 9

Network Modeling and Simulation: Summary

Here is the moment where many readers hope that the authors declare which M&S tool is best for them to use. Unfortunately, we can give no such advise. If not yet evident by the recurring themes of this book, the authors argue that there is no "best simulation tool." There is no silver bullet. There is no simulation tool that will ideally solve your exact problem. Conversely, there is no "worst simulation tool." So for those looking for the good, the bad, and the ugly, we apologize.

In this book, the authors have painstakingly avoided making any types of statements that could be construed as "product x" is better than "product y." However, in making a decision regarding what tool is best for your purpose it is understandable to seek the opinion of others on this subject. And indeed there are several comparative studies of simulation platforms in open literature. We refer the reader to [83–92] for additional information.

Each simulation tool that exists should be thought of as exactly that: a tool. Every tool has strengths and weaknesses depending on the intended use by the simulation designer. Commercial tools generally have a richer set of already-implemented protocol models compared with open source tools; however, open source tools have the advantage of being open source. They are free and there is a potentially large user community that is always contributing extensions to the tool free of charge. Commercial tools are generally not free, but offer services such as technical support and may have a richer set of data analysis and visualization tools. Some simulation tools provide better support of wireless networking technologies than other tools. Some simulation tools provide superior support for specific network appliances. Some tools provide better support to researchers for creation and testing of new protocols. Some simulation tools provide the ability to interact with live networks, supporting HITL concepts.

An Introduction to Network Modeling and Simulation for the Practicing Engineer, First Edition.
Jack Burbank, William Kasch, Jon Ward.
© 2011 Institute of Electrical and Electronics Engineers. Published 2011 by John Wiley & Sons, Inc.

The "best" tool to use is up to you, the network designer or developer. But before even choosing a tool, it behooves you to first carefully articulate your requirements. What are you trying to model? What is the required fidelity? How much detail do you need to include? It is also up to you, the network designer and developer, to properly verify and validate your models and simulations. It is then up to you to properly document and present your results so that they are both representative and reproducible.

It is up to you to decide whether a pure software-based simulation approach is the most appropriate, or whether a HITL approach is most advantageous. If a pure software approach is taken, is it most appropriate to have a single platform simulation or develop a distributed simulation? Only you, the designer, can decide. By taking time to understand the system you are wishing to simulate and by carefully articulating the requirements of the simulation activity, you can make informed decisions about the type of simulation capability you will utilize.

Remember, anyone can develop a simulation and present simulation results as if they are "ground truth." That does not make it useful. Just the opposite, it can be quite harmful. With due diligence and proper attention to the information presented in this book, both philosophical and practical, you can generate simulation results that are helpful to you, your organization, and the field of network design and development.

■■■■ REFERENCES

[1] W.T. Kasch, J.R. Ward, and J. Andrusenko, "Wireless Network Modeling and Simulation Tools for Designers and Developers," *IEEE Communications Magazine*, March 2009, Vol. 47, No. 3.

[2] D. Cavin et al., "On the Accuracy of MANET Simulators," Proceedings of the Workshop on Principles of Mobile Computing (POMC'02).

[3] I. Stojmenovic, "Simulations in Wireless Sensor and Ad Hoc Networks: Matching and Advancing Models, Metrics, and Solutions," *IEEE Communications Magazine*, December 2008, Vol. 46, No. 12.

[4] T.R. Andel, "On the Credibility of MANET Simulations," *IEEE Computer Society Computer*, July 2006, Vol. 39, Issue 7.

[5] T. Camp, J. Boleng, and V. Davies, "A Survey of Mobility Models for Ad Hoc Network Research," *Wireless Communication and Mobile Computing (WCMC)*: Special Issue on Mobile Ad Hoc Networking: Research, Trends and Applications, 2002, Vol. 2, No. 5, pp. 483–502.

[6] T. Issarivakul and E. Hossain, *Introduction to Network Simulator NS2*, Springer, 2008.

[7] T.S. Rappaport, *Wireless Communications: Principles and Practice*, 2nd Edition, Pearson Education, Upper Saddle River, NJ, 2002.

[8] G. Zhou, T. He, S. Krishnamurthy, and J.A. Stankovic, "Impact of Radio Irregularity on Wireless Sensor Networks," in Proceedings from MobiSys '04, pp. 125–138, June 2004.

[9] D. Kotz et al., "Experimental Evaluation of Wireless Simulation Assumptions," Proceedings of the 7th ACM International Symposium on Modeling, Analysis and Simulation of Wireless and Mobile Systems, 2004.

[10] M. Takai et al., "Effects of Wireless Physical Layer Modeling in Mobile Ad Hoc Networks," in Proceedings of the 2nd ACM International Symposium on Mobile Ad Hoc Networking and Computing, 2001.

[11] L. Devroye, *Non-Uniform Random Variate Distribution*, New York: Springer-Verlag, 1986.

[12] W. Press, S. Teukolsky, W. Vetterling, and B. Flannery, *Numerical Recipes: The Art of Scientific Computing*, Cambridge University Press, 2007.

An Introduction to Network Modeling and Simulation for the Practicing Engineer, First Edition.
Jack Burbank, William Kasch, Jon Ward.
© 2011 Institute of Electrical and Electronics Engineers. Published 2011 by John Wiley & Sons, Inc.

[13] The Network Simulator ns-2: Documentation, http://www.isi.edu/nsnam/ns/ns-documentation.html.

[14] J. Nuevo, A Comprehensible GloMoSim Tutorial, www.ccs.neu.edu/course/csg250/Glomosim/glomoman.pdf.

[15] OPNET Discrete Event Simulation Model Library, http://www.opnet.com/support/des_model_library/index.html.

[16] Scalable Network Technologies—QualNET Model Libraries, http://www.scalable-networks.com/products/libraries/models.php?lib=1 and http://www.scalable-networks.com/products/libraries/models.php?lib=9.

[17] "Part 11: Wireless LAN Medium Access Control (MAC) and Physical Layer (PHY) Specifications," IEEE 802.11–2007, 2007.

[18] R. Punnoose, P. Nikitin, and D. Stancil, "Efficient Simulation of Ricean Fading Within a Packet Simulator," in IEEE Vehicular Technology Conference, pp. 764–767, 2000.

[19] Additions to the OPNET Network Simulator to Handle Ricean and Rayleigh Fading, Academic OPNET Research and Educational Projects, http://www.ece.cmu.edu/wireless/arc_opnet.html.

[20] Cooperative for Scientific and Technical Research (COST) 231, Digital Mobile Radio Towards Future Generation Systems, Final Report, EUR18957, Chapter 4, 1999.

[21] R. Eichenlaub et al., "Fidelity at High Speed: Wireless InSite® Real Time Module™," Milcom 2008, November 2008.

[22] Remcom Wireless InSite, http://www.remcom.com/wireless-insite-features/.

[23] R. Eichenlaub, C. Valentine, S. Fast, and S. Albarano, "Fidelity at High Speed: Wireless InSite® Real Time Module™," IEEE Military Communications Conference (MILCOM) 2008.

[24] Token Ring Access Method and Physical Layer Specifications, IEEE Std 802.5-1985, 1985.

[25] L.Y. Tsui and O.M. Ulgen, "On Modeling Local Area Networks," 1988 Winter Simulation Conference Proceedings, Dec. 1988.

[26] Scalable Network Technologies—QualNET Model Libraries, http://www.scalable-networks.com/products/libraries/models.php?lib=0.

[27] CSMA/CD Illustrations, http://www.erg.abdn.ac.uk/users/gorry/course/lan-pages/csma-cd.html.

[28] M. Abdelhafez and G. Riley, "Revisiting Effects of Detail in Ethernet Simulations," 9th Communications and Networking Simulation Symposium (CNS 2006), 2006.

[29] A. Hussain, A. Kapoor, and J. Heidemann, "The Effect of Detail on Ethernet Simulation," in Proceedings of the ACM Workshop on Parallel and Distributed Simulation, 2004.

[30] IEEE Standard 802.3, Carrier Sense Multiple Access with Collision Detection (CSMA/CD) Access Method and Physical Layer Specifications, 2008.

[31] M.C. Jeruchim, P. Balaban, and K.S. Shanmugan, *Simulation of Communication Systems—Modeling, Methodology, and Techniques*, Kluwer Academic/Plenum Publishers, 2000.

[32] M. Jeruchim, "Techniques for Estimating the Bit Error Rate in the Simulation of Digital Communication Systems," *IEEE Journal on Selected Areas in Communications*, Vol. 2, Issue 1, January 1984.

[33] J. Rinas, "Reference Curves for Linear Modulation Schemes," ANT Department of Communications Engineering, http://www.ant.uni-bremen.de/whomes/rinas/dfsim/linearmodreference.html.

[34] A. Chandra, V. Gummalla, and J. Limb, "Wireless Medium Access Control Protocols," *IEEE Communications Surveys*, http://www.comsoc.org/pubs/surveys, Second Quarter, 2000.

[35] L.F. Perrone, "Modeling and Simulation Best Practices for Wireless Ad Hoc Networks," in Proceedings of the 2003 Winter Simulation Conference, Vol. 1, December 2003.

[36] J. Heidemann et al., "Effects of Detail in Wireless Network Simulation," in Proceedings of the SCS Multiconference on Distributed Simulation, 2001.

[37] "Part 16: Air Interface for Fixed Broadband Wireless Access Systems," IEEE 802.16–2044, 2004.

[38] C. Wullems, K. Tham, J. Smith, and M. Looi, "A Trivial Denial of Service Attack on IEEE 802.11 Direct Sequence Spread Spectrum Wireless LANs," Wireless Telecommunications Symposium, 2004.

[39] S. Kumar, V. Raghavan, and J. Deng, "Medium Access Control Protocols for Ad Hoc Wireless Networks: A Survey," *ElsevierAd Hoc Networks Journal*, 2004.

[40] R. Jurdak, C. Videira Lopes, and P. Baldi, "A Survey, Classification and Comparative Analysis of Medium Access Control Protocols for Ad Hoc Networks," *IEEE Communications Surveys & Tutorials*, First Quarter 2004, Vol. 6, No. 1.

[41] S. Mehta et al., "A Case Study of Networks Simulation Tools for Wireless Networks," Third Asia International Conference on Modeling & Simulation, 2009 (AMS '09), May 2009.

[42] A. Tanenbaum, *Computer Networks*, 4th Edition, Prentice-Hall, Upper Saddle River, NJ, 2003.

[43] G. Anastasi and L. Lenzini, "QoS Provided by the IEEE 802.11 Wireless LAN to Advanced Data Applications: A Simulation Analysis," *Wireless Nets*, 2000, Vol. 6, No. 2, pp. 99–108.

[44] G. Bianchi, "Performance Analysis of the IEEE 802.11 Distributed Coordination Function," *IEEE JSAC*, 2000, Vol. 18, No. 3, pp. 535–547.

[45] G. Zhou, T. He, S. Krishnamurthy, and J.A. Stankovic, "Impact of Radio Irregularity on Wireless Sensor Networks," in Proceedings from MobiSys '04, pp. 125–138, June 2004.

[46] M.S. Gast, *802.11 Wireless Networks: The Definitive Guide*, O'Reilly & Associates, April 2002.

[47] WiMAX Forum Website, http://www.wimaxforum.org/.

[48] T. Cooklev, *Wireless Communications Standards: A Study of IEEE 802.11, 802.15, and 802.16*, IEEE Press, 2004.

[49] WiMAX MAC Extension for NS-2, http://www.lrc.ic.unicamp.br/wimax_ns2/.

[50] I. Stojmenovic, "Simulations in Wireless Sensor and Ad Hoc Networks: Matching and Advancing Models, Metrics, and Solutions," *IEEE Communications Magazine*, December 2008, Vol. 46, No. 12.

[51] J. Heidemann et al., "Expanding Confidence in Network Simulations," *IEEE Network*. 2001, September–October, Vol. 15, No. 5.

[52] D. Kotz et al., "Experimental Evaluation of Wireless Simulation Assumptions," Proceedings of the 7th ACM International Symposium on Modeling, Analysis and Simulation of Wireless and Mobile Systems, 2004.

[53] J. Heideman et al., "Effects of Detail in Wireless Network Simulation," in Proceedings of the SCS Multiconference on Distributed Simulation, 2001.

[54] V. Paxon and S. Floyd, "Why We Don't Know How to Simulate the Internet," in Proceedings of the 1997 Winter Simulation Conference, pp. 1037–1044, 1997.

[55] G. Riley and M. Ammar, "Simulating Large Networks—How Big Is Big Enough?" Conference on Grand Challenges for Modeling and Simulation, January 2002.

[56] D. Xu et al., "Enabling Large-Scale Multicast Simulation by Reducing Memory Requirements," in Proceedings of the Seventeenth Workshop on Parallel and Distributed Simulation, 2003.

[57] P. Huang, D. Estrin, and J. Heideman, "Enabling Large-Scale Simulations: Selective Abstraction Approach to the Study of Multicast Protocols," in Proceedings of the International Symposium of Modeling, Analysis, and Simulation of Computer and Telecommunication Systems, July 1998.

[58] G. Riley, M. Ammar, and R. Fujimoto, "Stateless Routing in Network Simulations," in Proceedings of the Eighth International Symposium on Modeling, Analysis, and Simulation of Computer and Telecommunications Systems, August 2000.

[59] D. Xu, G. Riley, M. Ammar, and R. Fujimoto, "Enabling Large-Scale Multicast Simulation by Reducing Memory Requirements," in Proceedings of Workshop on Parallel and Distributed Simulation, 2003.

[60] B. Divecha, A. Abraham, C. Grosnan, and S. Sanyal, "Impact of Node Mobility on MANET Routing Protocols Models," *Journal of Digital Information Management*, February 2007.

[61] G. Lin, G. Noubir, and R. Rajaraman, "Mobility Models for Ad-Hoc Network Simulation," in Proceedings of IEEE INFOCOM 2004, Vol. 1, pp. 7–11, 2004.

[62] T. Camp, J. Boleng, and V. Davies, "A Survey of Mobility Models for Ad-hoc Networks," *Special Issue on Mobile Ad-Hoc Networking: Research, Trends, and Applications*, 2002, Vol. 2, No. 5, pp. 483–502.

[63] F. Bai and A. Helmy, "The IMPORTANT Framework for Analyzing and Modeling the Impact of Mobility in Wireless Ad-Hoc Networks," in *Wireless Ad-Hoc and Sensor Networks*, Kluwer Academic Publishers, 2004.

[64] F. Bai, N. Sadagopan, and A. Helmy, "User Manual for IMPORTANT Mobility Tool Generators in NS-2 Simulator," http://nile.cise.ufl.edu/important/mobility-user-manual.pdf, release date February 2004.

[65] The NSWEB Traffic Generator for NS-2.29, http://www.net.t-labs.tu-berlin.de/~joerg/.

[66] The PackMime-HTTP Traffic Generator for NS-2, http://www.dirt.cs.unc.edu/packmime/.

[67] TMix: A Tool for Generating Realistic TCP Application Workloads in NS-2, http://ccr.sigcomm.org/online/?q=node/50.

[68] Harpoon: A Flow-Level Traffic Generator, http://pages.cs.wisc.edu/~jsommers/harpoon/.

[69] W.T. Kasch and J.L. Burbank, "The Evaluation of Wireless Networking through ACTION," IEEE Military Communications Conference, October 2005.

[70] R. Martinez et al., "Hardware and Software-in-the-Loop Techniques using the OPNET Modeling Tool for JTRS Developmental Testing," IEEE Military Communications Conference, October 2003.

[71] L. Carter et al., "A Hardware-in-the-Loop Network Simulator for Analysis and Evaluation of Large-Scale Military Wireless Communication Systems," IEEE Military Communications Conference, November 2008.

[72] SITL Module Brochure, Opnet Technologies, Inc., 2006.

[73] S. Doshi et al., "JMEE: A Scalable Framework for JTRS Waveform Modeling and Evaluation," IEEE Military Communications Conference, November, 2008.

[74] IEEE Standard 1516-2000, IEEE Standard for Modeling and Simulation (M&S) High Level Architecture (HLA)—Framework and Rules.

[75] IEEE Standard 1516.1-2000, IEEE Standard for Modeling and Simulation (M&S) High Level Architecture (HLA)—Federate Interface Specification.

[76] IEEE Standard 1516.2-2000, IEEE Standard for Modeling and Simulation (M&S) High Level Architecture (HLA)—Object Model Template (OMT) Specification.

[77] "IEEE Recommended Practice for High Level Architecture (HLA) Federation Development and Execution Process (FEDEP)," IEEE Standard 1516.3-2003, Approved March 20, 2003, Copyright 2003, IEEE.

[78] D. Rhodes et al., "Scalable Urban Network Simulation (SUNS)," IEEE Military Communications Conference, October 2007.

[79] B. Liu et al., "VGSim: An Integrated Networking and Microscopic Vehicular Mobility Simulation Platform," *IEEE Communications Magazine*, May 2009.

[80] "Verification, Validation, and Accreditation (VV&A) Recommended Practices Guide," Office of the Director of Defense Research and Engineering Defense Modeling and Simulation Office, November 1996.

[81] "Use of Modeling and Simulation (M&S) in Operational Testing," COMOPTEVFORINST 5000.1A, U.S. Navy Commander Operational Test and Evaluation Force, September 9, 2004.

[82] http://toilers.mines.edu/Public/NsInspect.

[83] G.A. Di Caro, "Analysis of Simulation Environments for Mobile Ad Hoc Networks," Technical Report No. IDSIA-24-03, December 2003.

[84] G. Flores Lucio et al., "OPNET Modeler and NS-2: Comparing the Accuracy of Network Simulators for Packet-Level Analysis Using a Network Testbed."

[85] T. Watteyne, "Using Existing Network Simulators for Power-Aware Self-Organizing Wireless Sensor Network Protocols," INRIA No 6020, September 2006.

[86] D.M. Nicol, "Comparison of Network Simulators Revisited," http://www.ssfnet.org/Exchange/gallery/dumbbell/dumbbell-performance-May02.pdf.

[87] J. Lessmann, P. Janacik, and D. Orfanus, "Comparative Study of Wireless Network Simulators," Seventh International Conference on Networking, April 2008.

[88] D. Orfanus, J. Lessmann, P. Janacik, and L. Lachev, "Performance of Wireless Network Simulators: A Case Study," in Proceedings of the 3rd ACM Workshop

on Performance Monitoring and Measurement of Heterogenerous Wireless and Wired Networks.

[89] A. Brown and M. Kolbert, "Tools for Peer-to-Peer Network Simulation," draft-irtf-p2prf-core-simulators-00.txt.

[90] F. Kargl and E. Schoch, "Simulation of MANETs: A Qualitative Comparison Between JiST/SWANS and NS-2," in MobiEval 2007: Proceedings of the 1st International Workshop on System Evaluation for Mobile Platforms, pp. 41–46, 2007.

[91] Y. Xue et al., "Performance Evaluation of NS-2 Simulator for Wireless Sensor Networks," in Proceedings of the 2007 Canadian Conference on Electrical and Computer Engineering, pp. 1372–1375, April 2007.

[92] P. Garrido, M. Malumbres, and C. Calafate, "NS-2 vs. OPNET: A Comparative Study of the IEEE 802.11e Technology on MANET Environments," in Proceedings of the First International Conference on Simulation Tools and Techniques for Communications, Networks, and Systems, 2008.

[93] Rec. ITU-R M.1225 1, *Guidelines for Evaluation of Radio Transmission Technologies for IMT-2000*, 1997.

[94] A.F. Molisch, *Wireless Communications*, Wiley, 2005, and associated web page http://www.wiley.com/legacy/wileychi/molisch/supp/appendices/Chapter_7_Appendices.pdf.

[95] European Cooperative in the Field of Science and Technical Research EURO-COST 231, "Urban Transmission Loss Models for Mobile Radio in the 900- and 1,800 MHz Bands (Revision 2)," COST 231 TD(90)119 Rev. 2, The Hague, The Netherlands, September 1991, available at http://www.lx.it.pt/cost231/final_report.htm.

[96] WiMAX Forum Mobile Radio Conformance Tests, "DRAFT-T25-002-R010v03-B Working Group Approved Revision (2009-07-06)," July 2009.

[97] Scalable Network Technologies—QualNET Advanced Wireless Library Model (WiMAX—IEEE 802.16d/e), www.scalable-networks.com/.../qualnet-advanced-wirelesswimax-library-data-sheet/.

[98] OPNET WiMAX (802.16) Specialized Model for OPNET Modeler Wireless Suite: http://www.opnet.com/WiMAX/.

[99] NS2 WiMAX Model version 2.6 from WiMAX Forum Member's Website: www.wimaxforum.org.

[100] Azimuth Systems ACE 400WB MIMO Channel Emulator, http://www.azimuthsystems.com/Collateral/Documents/Common/PB_ACE400wb_0408_v9_sql.pdf.

[101] Elektrobit EB Propsim C8—Multi-Channel Emulator, http://www.elektrobit.com/what_we_deliver/wireless_communications_tools/products/eb_propsim_c8.

[102] MATLAB and Simulink, http://www.mathworks.com/.

[103] MATLAB Central: Open exchange for the MATBLAB and Simulink User Community, http://www.mathworks.com/matlabcentral/index.html.

[104] H. MacLeod, C. Loadman, and Z. Chen, "Experimental Studies of the 2.4-GHz ISM Wireless Indoor Channel," in Proceedings of the 3rd Communication Networks and Services Research Conference (CNSR), p. 63, 2005.

[105] "WiMAX Forum System Level Simulator NS-2 MAC+PHY Add-On for WiMAX (IEEE 802.16)," Version 2.6 (Beta), March 20, 2009. Draft.

[106] L.W. Couch, II, *Digital and Analog Communication Systems*, 6th Edition, Prentice Hall, 2001.

[107] D. Reddy and G. Riley, "Measurement Based Physical Layer Modeling for Wireless Network Simulations," in *Modeling, Analysis, and Simulation of Computer and Telecommunication Systems, 2007. MASCOTS '07. 15th International Symposium*, pp. 46–53, October 24–26, 2007.

[108] W. Wang, H. Sharif, M. Hempel, T. Zhou, P. Mahasukhon, and T. Ma, "Implementation and Performance Evaluation of a Complete, Accurate, Versatile and Realistic Simulation Model for Mobile WiMAX in NS-2," *2010 IEEE International Communications Conference (ICC)*, pp. 1–5, May 23–27, 2010.

[109] NS-2 Wireless Package Tutorial, http://www.isi.edu/nsnam/ns/tutorial/.

[110] NS-2 Download Location, http://www.isi.edu/nsnam/ns/.

[111] NS-2 Trace File Formats, http://nsnam.isi.edu/nsnam/index.php/NS-2_Trace_Formats#New_Wireless_Trace_Formats.

[112] Summary of IEEE 802.11 parameters for NS-2, http://www.joshuarobinson.net/docs/ns-802_11b.html.

[113] The Network Simulator NS-2 NIST add-on IEEE 802.16 Model (MAC+PHY), www.nist.gov/itl/antd/emntg/upload/wimax_module.pdf.

[114] A. Saleh and R. Valenzuela, "A Statistical Model for Indoor Multipath Propagation," *IEEE Journal on Selected Areas in Communications*, February 1987, Vol. 5, No. 2, pp. 128–137.

[115] E. Silva and G.A. Carrijo, "A Vectorial Analysis of the Two-Ray Model," *Communications Systems, 2004. ICCS 2004. The Ninth International Conference*, pp. 607–611, September 2004.

[116] B. Sklar, "Rayleigh Fading Channels in Mobile Digital Communications Systems—Part I: Characterization," *IEEE Communications Magazine*, July 1997.

[117] N. Cooper and Natarajan Meghanathan, "Impact of Mobility Models on Multipath Routing in Mobile Ad-Hoc Networks," *International Journal of Computer Networks and Communications (IJCNC)*, January 2010, Vol. 2, No. 1.

[118] Mohd Izuan Mohd Saad, "Performance Analysis of Random-Based Mobility Models in MANET Routing Protocol," *European Journal of Scientific Research*, 2009, Vol. 32, No. 4, pp. 444–454.

[119] C. Shete et al., "Analysis of the Effects of Mobility and Node Density on the Grid Location Service in Ad Hoc Networks," *IEEE Conference on Communications (ICC)*, June 2004.

[120] V. Loscri, E. Natalizio, and C. Costanzo, "Simulations of the Impact of Controlled Mobility for Routing Protocols," *EURASIP Journal on Wireless Communications and Networking*, April 2010, Vol. 2010, No. 7.

[121] H. Heffes and D.M. Lucantoni "A Markov Modulated Characterization of Packetized Voice and Data Traffic and Related Statistical Multiplexer Performance," *IEEE Journal on Selected Areas in Communications*, September 1986, Vol. 4, No. 6, pp. 856–868.

[122] R. Jain, S.A. Routhier, "Packet Trains—Measurements and a New Model for Computer Network Traffic," *IEEE Journal on Selected Areas in Communications*, September 1986, Vol. 4, No. 6, pp. 986–995.

[123] W.E. Leland, M.S. Taqqu, W. Willinger, and D.V. Wilson, "On the Self-Similar Nature of Ethernet Traffic (Extended Version)," *IEEE/ACM Trans. Netw.* February 1994, Vol. 2, No. 1, pp. 1–15.

[124] W. Willinger, "The Discovery of Self-Similar Traffic," in *Performance Evaluation: Origins and Directions*. G. Haring, C. Lindemann, and M. Reiser, Eds. Lecture Notes in Computer Science, Vol. 1769. Springer-Verlag, London, pp. 513–527.

[125] A. Erramilli, M. Roughan, D. Veitch, and W. Willinger. "Self-Similar Traffic and Network Dynamics," in Proceedings of the IEEE, 2002, Vol. 90, No. 5.

[126] K. Salah and A. Alkhoraidly, "An OPNET-Based Simulation Approach for Deployment VoIP," *International Journal of Network Management*, May 2006, Vol. 16, No. 3, pp. 159–183.

[127] S. Rakitin, *Software Verification and Validation for Practictioners and Managers*, 2nd Edition, Artech Print, August 2001.

[128] W. Oberkampf and C. Roy, *Verification and Validation in Scientific Computing*, Cambridge University Press, November 2010.

[129] M. Fisher, *Software Verification and Validation: An Engineering and Scientific Approach*, Springer, November 2010.

[130] High-Level Architecture Rules, Version 1.3, U.S. Department of Defense, April 1998.

[131] High Level Architecture Interface Specification, Version 1.3, U.S. Department of Defense, April 1998.

[132] High-Level Architecture Object Model Template (OMT) Specification, Version 1.3, U.S. Department of Defense, April 1998.

[133] Xilinx System Generator for DSP Simulink Blockset, http://www.mathworks. com/products/connections/product_detail/product_35567.html.

[134] LabVIEW FPGA, http://www.ni.com/fpga/.

An Introduction to Network Modeling and Simulation for the Practicing Engineer, First Edition.
Jack Burbank, William Kasch, Jon Ward.
© 2011 Institute of Electrical and Electronics Engineers. Published 2011 by John Wiley & Sons, Inc.

Printed in the United States
By Bookmasters